# Public Transport Planning with Smart Card Data

# Public Transport Planning with Smart Card Data

*Editors*

**Fumitaka Kurauchi**
Department of Civil Engineering
Faculty of Engineering
Gifu University
Gifu, Japan

**Jan-Dirk Schmöcker**
Department of Urban Management
Graduate School of Engineering
Kyoto University
Kyoto, Japan

CRC Press
Taylor & Francis Group
Boca Raton London New York

CRC Press is an imprint of the
Taylor & Francis Group, an **informa** business
A SCIENCE PUBLISHERS BOOK

CRC Press
Taylor & Francis Group
6000 Broken Sound Parkway NW, Suite 300
Boca Raton, FL 33487-2742



Printed on acid-free paper
Version Date: 20160725

International Standard Book Number-13: 978-1-4987-2658-0 (Hardback)



**Visit the Taylor & Francis Web site at**
**http://www.taylorandfrancis.com**

**and the CRC Press Web site at**
**http://www.crcpress.com**

# Preface

Collecting fares through "smart cards" is becoming standard in most advanced public transport networks of major cities around the world. Using such cards has advantages for users as well as operators. Whereas for travellers smart cards are mainly increasing convenience, operators value in particular the reduced money handling fees. Smart cards further make it easier to integrate the fare systems of several operators within a city and to split the revenues. The electronic tickets also make it easier to create complex fare systems (time and space differentiated prices) and to give incentives to frequent or irregular travellers. Less utilized though appear to be the behavioural data collected through smart card data. The records, even if anonymous, allow for a much better understanding of passengers' travel behaviour as various literature has begun to demonstrate. This information can be used for better service planning.

This book handles three major topics; how passenger behaviour can be estimated using smart card data, how smart card data can be combined with other trip databases, and how the public transport service level can be better evaluated if smart card data are available. The book discusses theory as well as applications from cities around the world.

September 2016

**Fumitaka Kurauchi**
**Jan-Dirk Schmöcker**

# Contents

Chapter **1**

# An Overview on Opportunities and Challenges of Smart Card Data Analysis

**J.-D. Schmöcker[1,*], F. Kurauchi[2] and H. Shimamoto[3]**

**ABSTRACT**

In this chapter, an overview on opportunities and challenges for the use of smart card data for public transport planning is provided. As an introduction to the topic examples of customer services that have become feasible due to smart cards are discussed. These include smart card as a general payment method for a wide range of services, pricing caps as well as "loyalty points". For operators, smart cards provide opportunities such as revised fare structure. The focus of this chapter and this book in general is on the benefits that emerge through better understanding of customer behavioural patterns for short and longer term service planning. This chapter also points out that in practice smart card data are though not yet as much used as one might expect given these opportunities. As explanation for this challenges connected to big data issues, privacy and missing information are discussed. The chapter concludes by providing an overview on the contributions in this book.

## 1. INTRODUCTION

Automatic Fare Collection through "smart cards" is becoming a standard in most advanced public transport networks of major cities around the world. Using such cards has an advantage for users as well as operators. Whereas smart cards are mainly increasing convenience for travellers, operators value in particular the reduced money handling fees. Smart cards further make it easier to integrate the fare systems of several operators within a city and to split the revenues.

[1] Department of Urban Management, Graduate School of Engineering, Kyoto University, Japan. Email: schmoecker@trans.kuciv.kyoto-u.ac.jp

[2] Department of Civil Engineering, Gifu University, Japan. Email: kurauchi@gifu-u.ac.jp

[3] Department of Civil and Environmental Engineering, University of Miyazaki, Japan. Email: shimamoto@cc.miyazaki-u.ac.jp

* Corresponding author

These are the primary reasons that led in many cities to invest in the introduction of smart card systems. The focus of this book is though the secondary benefits that are obtained through smart card data. Smart card data are increasingly recognised as a rich data source to better understand demand patterns of passengers. As this book will discuss, origin-destination matrices, routes and activities all can be inferred from this data. Furthermore, smart card data can be used partly as replacement of other data sources to collect evaluation measures of the service quality. That is, the time and the location stamps of the records allow the operator to measure, for example, actual versus the scheduled arrivals of the buses.

Before discussing the analysis options in detail the following section will give an overview on the spread of smart card systems across the world, including the differences in the collected data. Recognizing these differences is not only important to understand the analysis potential but also to understand the challenges an analyst faces. These challenges together with a discussion on actual usage of smart card data in practice is the topic of Section 4.

Section 5 then provides an overview on the contents of the following chapters in the book. The primary purpose of the book is to provide an overview on smart card data analysis opportunities and how challenges are overcome. Evidently, considering that the literature on smart card data is rapidly growing, the book does not claim completeness. The section will hence briefly discuss further data analysis options and examples which could be perceived as important but missing in this book before concluding.

## 2. SMART CARD SYSTEMS AND DATA FEATURES

The numbers of smart cards are increasing year by year, for example Wikipedia lists more than 350 smart card systems all over the world covering all continents. As this book focuses on smart card systems that have their primary application payment for public transport, one needs to recognise that smart cards are in use for a wider range of applications. An important development is therefore the integration of different applications into smart card systems.

Through the worldwide spread of smart cards, international standardization, which define the signal frequency and the data transmission speed, has progressed. For the contactless cards there are several standards that cover the lower levels of interface between cards and terminals and mainly three types of standard, referred to as Type A, Type B and FeliCa, are widely prevalent. For transit smart cards, either Type-A or FeliCa systems are adopted. Type-A systems are common all over the world since they could be introduced with low cost. The biggest advantage of the FeliCa system is the faster transmission speed. Due to this feature, FeliCa system cards prevail in many transit companies in Japan where it is essential to handle large amount of passengers in short time during the rush hours. For further detailed criteria of these standards, readers can refer to Pelletier et al. (2011). Table 1 shows information on the selection of noteworthy major smart cards that are issued mainly for the purpose of transportation fare collection. For users (and data analysts) the increasing standardization further means that not only the arrangement of same card usage for different operators becomes easier but also the usage of the same card in different cities. For example, in Japan since 2013 most of the smart cards from major public operators can be used across the country. The Netherlands is one of the first countries where a single smart card can be used throughout the country for local as well as long distance travel.

The important aspect for data analysis and transport demand management possibilities is whether the transactions are pre-paid (debit) or post-paid (credit). Although most of the smart card systems adopt the pre-paid system, an increasing number also offer post-payment systems, mostly not in replacement but in addition to pre-paid ones. This means, that, similar to credit cards, the total transportation fares accumulated over a month will debit from the bank account next month. The drawback of the post-payment system for the user is that it requires personal details and an application for qualification to get the cards. This means that it often takes a considerable amount of time until the cards are issued. However, the post-paid system cards also have some merits for the users. First of all, since the bank debits the fare later from the account, users do not have to worry about the remaining money on the card. Secondly, with personalized post-payment cards, loyalty schemes are more widely spread. One example is the "PiTaPa" card, which could be used for fare payment on most of the private trains and bus companies in the Kansai region of Japan. Operators utilizing PiTaPA offer different amount of discounts per journey and some set an upper limit for the fare-to-be paid for pre-registered origins and destinations by the users. For other (not pre-registered) journeys PiTaPa also offers discount based on how much fare the users have paid or how often the users have used PiTaPA for public transport during the previous month. Furthermore, some of the transit companies in Japan give points for the users based on the boarding history as well as the shopping history at the designated shops. In Chapter 7 this is further discussed with the help of an example of Shizutetsu Railway Co., Ltd., a private rail operator in Shizuoka, Japan. The cardholders can use these points for fare or shopping discounts in stores associated with the transport operator. Therefore, for demand management, in general post-paid systems are preferable. For the data analyst post-paid systems further mean that travel data and socio-demographic data required for registration can be obtained, though obviously privacy issues are a major concern for this.

Table 1 includes some additional observations on selected smart cards that appear noteworthy to us: The Octopus card was one of the early card schemes not only for transport but also in general promoting the usage of

**Table 1.** Information on selected smart card systems

| Name of Card | City and Country | Year of Introduction | Noteworthy Points (but not necessarily unique features of these cards) |
|---|---|---|---|
| Octopus Card | Hong Kong, China | 1997 | Various added functions, including payment at international chains such as Starbucks or McDonald's. Currently replacement of 1st generation cards: 2nd generation cards allow, among others, online payment. |
| Suica | Various metropolitan areas in Japan | 2001 | The fare calculation is by one yen unit with the smart card whereas the fare calculation for paper-based tickets is by ten yen units. Mutual use of other smart cards such as ICOCA or PASMO. |
| Oyster Card | London, UK | 2003 | Paying by smart card is much cheaper than paper ticket; "daily cap" and "weekly caps" are implemented on smart cards. |
| T-money | Various metropolitan areas in Korea | 2004 | Over 100 million cards (accumulated) are allotting by now (Korea smart card, 2016). The system is also supplied to operators outside Korea. Chapter 3 shows an application of analysis with T-Money data from Seoul. |
| OV-Chip Card | Nationwide in the Netherlands | 2005 (Rotterdam only) | Can be used for almost all public transport in the Netherlands, including local and long distance travel (see Chapter 12). |
| LuLuCa | Shizuoka, Japan | 2006 | Extensive loyalty point scheme to encourage usage of card for transit as well as for shopping (see Chapter 7). |
| Bip! Card | Santiago, Chile | 2007 | Bip! Card is the only allowed payment method on buses. (see Chapters 2 and 9) |

the card for different purposes, which is also included in the etymology of the card's name. Nowadays, the card could be used for a variety of shopping including online purchases.

Several operators have also been promoting the uptake of smart cards by providing cheaper fares compared to paper tickets. Noteworthy are the discounts provided in London, where paper tickets can be priced double compared to the payment by Oyster card. In Japan, generally no discounts are given for the usage of smart cards. Recently though, due to an increase in the VAT, there are small price differences between paper tickets and payments by smart cards. The increase in fares due to VAT raise is reflected accurate to 1 Yen for smart cards where paper tickets are rounded to the nearest 10 Yen. Such minor price differences are though unlikely to have an impact on travel decisions. More important might be the effect of "daily caps" or, recently, "weekly caps" that have been applied in London. These caps mean that the user does not have to decide in the morning or the beginning of the week anymore whether it will be worth purchasing a daily or weekly pass. Instead the traveller has the guarantee that the smart card will stop charging the user if the equivalent prices of a daily or weekly pass has been accumulated through single fares. In how far this scheme has any impact on behaviour is not yet known to our knowledge. Finally, it should be noted that in some cities, such as Santiago, it is compulsory for users to get a smart card as cash payment on some modes of transport is not possible anymore.

## 3. ANALYSIS CHALLENGES

As the smart cards are widely spread one might expect that their historical data records have also been exploited heavily for transportation planning. This appears tough for many operators not yet to be the case. Imai et al. (2012) conducted a survey among 66 Japanese operators asking them about the purposes they use the smart card data for. The results are shown in Figure 1. One can see that many operators do not utilize the smart data card for transport planning purposes at all. From those who use the data, the majority uses them only for some simple collective analysis or for reporting purposes. This situation is probably not unique to Japan and also in other countries it will be often only large, or a few innovative, transport operators that have enough resources to dedicate themselves to the analysis of the vast amount of data that they obtain from the smart cards.

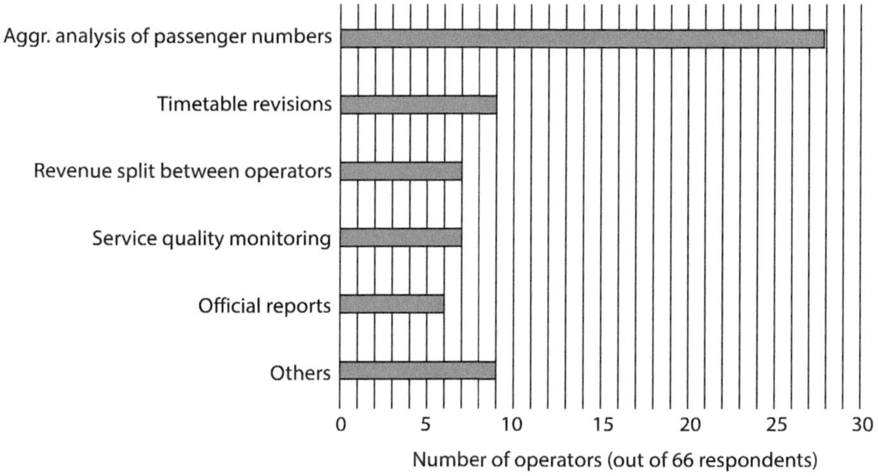


**Fig. 1.** Usage of smart card data by operators in Japan according to a survey in 2012

*Source*: Table adjusted from Imai 2012.

A main reason for this situation is that, although most would agree that the potential information to be derived from the data is useful, there are also several challenges to be overcome before the data become in fact useful. A list of data potentials and challenges is given in Table 2. The importance/benefits of the first two points (data at lower cost, aggregate performance statistics) will be fairly obvious to most operators. The latter two points on more detailed information about travellers will especially help providers to develop strategies to better target the services. This discussion continues in the next section awhereas the focus in this section is on the challenges.

The first challenge, the representativeness of population from the smart card sample, may not be a significant problem anymore in many cities since

**Table 2.** Potential and challenges of smart card data that motivate this book

| Advantages/Potential | Disadvantages/Challenges |
|---|---|
| • To get large amount of data on passengers' behaviour with lower cost | • Representativeness of population is not guaranteed |
| • To analyse aggregate behaviour including "dynamic aspects" | • Big data issues |
| • To analyse data on personal level to understand variation in behaviour | • Privacy and contractual issues |
| • To match data with other information (e.g., purchase history during the trip) | • Missing information |

the rate of payment by smart cards is increasing year by year. Nevertheless, operators need to be aware that in particular irregular users might be under-represented in the smart card data sample.

Connected to the increasing data size are though also "big data issues". Since smart cards collect daily passenger behaviour continuously, the data size may become so large that it is sometimes difficult to handle. Smart card data can therefore be regarded as one type of 'big data'. A major difference to traditional data analysis is that 'big data' often provide information on nearly the whole system population. In traditional data analysis, a 'hypothesis' should be first set and sampling should be carried out based on this hypothesis. Then the population characteristics assessment is done by the sample data and the hypothesis is tested. In contrast in big data analysis such a sampling strategy is not needed any more. What instead becomes important in big data analysis is how relevant samples are picked up and how important information will be extracted from the data. Statistical methods such as factor analysis and/or clustering analysis are often adopted to understand the sample characteristics, but the procedure is far more difficult considering the data size. Also, one should recognise that when using big data, it becomes too easy to reject the null hypothesis of no statistical significance as discussed in Harding 2013. Therefore, special consideration might be necessary in handling big data.

The second challenge, privacy issues, occurs in handling smart card data since the cards can contain private information, including monetary information, especially if it is a post-payment card. This makes it often difficult to get access to smart card data and/or to develop analysis methodologies that remain data confidentiality. Ideally, a universal rule in utilizing smart card data in public transport service management and evaluation should be discussed, though this will be difficult given different law constraints in different countries. Similar to privacy rules, there is often a contract that data must not be given to others to protect a possible deficiency. Such a contract is active especially when different companies are sharing the same card such as, in Japan, PASMO in Tokyo metropolitan area and the PiTaPa card in the Kansai area.

Another common challenge encountered by analysts is missing information. This could be due to above-mentioned privacy regulations, due to missing records, or simply because they are not recorded with smart card data. In particular for pre-paid smart cards there are usually few or no socio-demographic information recorded. Chapters 3 and 5 in this book will discuss some probabilistic approaches to overcome such challenges. Further important information may not be recorded due to the fare system. For example, bus companies that adopt flat fare systems only record either the boarding or alighting bus stop since there is no need for passengers to tap in and out. Also, in subways where ticketing gates at stations are common among lines, information on the routes taken by travellers may not be recorded as will be discussed more in Chapter 4. In summary, though some of these missing information constraints can be overcome, in many cases more analysis processes are often required before the data deliver some useful information.

## 4. CATEGORIZATION OF POTENTIAL ANALYSIS USING SMART CARD DATA

Despite all these challenges, when properly analysed, the smart card data can be a very powerful tool, for service management as shown in the contributions in this book. In their review on the potential for smart card data Pelletier et al. (2011) noted that smart card data can be used for strategic-level, tactic-level and long-term planning which they define as:

***Strategic-level studies***: Long-term planning. An understanding of tendency of passengers' behaviour for long-term planning such as demand forecasting and marketing. An example of the analysis from this level is classification of travellers.

***Tactical-level studies***: Service adjustments and network development. Determine patterns in travel behaviour to adjust service frequency and route. An example of the analysis from this level is transfer journey.

***Operational-level studies***: Ridership statistics and performance indicators. An understanding of detail in passengers' behaviour to measure the performance indicator. An example of the analysis from this level is schedule adherence.

One might further extend this classification as in Table 3.

If smart card data are aggregated, one can get knowledge and create graphs to illustrate details of travellers' demand for strategic planning as shown in Chapter 9 or in various literature such as Jang (2010) with data from Seoul. Without smart card data these details are gained from boarding and alighting count surveys with great effort. Moreover, as mentioned before, one of the advantages of the use of smart card data is that it is possible to track individual behaviour. Therefore, from the analysis of the individual demand data, one can infer popular transfer

**Table 3.** Possible analysis using smart card data

| Extracted Data/ Level | Space Dimension | Level of Analysis | Examples for Use by Operators |
|---|---|---|---|
| Demand, aggregated | Stop | Strategic | Directly for service planning. |
| | Line | | |
| | Network | | |
| Demand, individual cross-sectional data | Route | Tactical | Design services so that it allows for choice flexibility ("hyper-paths"). |
| | OD patterns | | Minimize transfers and journey times, distribution by time of day. |
| | Trip chains, Journeys[1] | | Where to offer transfer information and waiting facilities. |
| Demand, individual panel data (card ID could be tracked over time) | Route | Tactical | Estimation of demand variation over time. |
| | OD patterns | | Allows distinguishing "white noise" from explainable demand variation for capacity planning. |
| | | | Prediction of possible consequences of service disruption and infrastructure investments. |
| | Trip chains, Journeys | Strategic | Service adjustments to user travel needs. |
| Supply[2] | Stop | Operational | Evaluation criteria: Regularity, waiting time. |
| | Route | | Evaluation criteria: km operated, schedule adherence, "bunching". |
| | Network | | As for routes, plus, e.g., knock-on effects of delays between routes. |

*Notes:*

[1] Need alighting data, in some systems not available, might be inferable, see Chapter 2.

[2] In some systems such data can be directly extracted from smart card data, in others, like London, a separate data system (ibus) provides this data (see Chapter 8 where Singapore bus departure times are estimated from smart cards).

points, which is essential information for providing transfer facilities or even for long-term bus network planning, (Jang 2010). Furthermore, if one analyses individual time series data, it is possible to capture the day-to-day variation of travellers' demand or their chosen route (set). It is suggested that one contribution of this is for better understanding of network reliability. Although many advanced network models have been proposed to deal with demand uncertainty, most of these assume that the demand or route choice probability follow a certain (simple) probabilistic distribution due to difficulties in obtaining good panel data. Instead, with smart card data it is possible to detect such distributions and/or to distinguish traveller groups according to their demand variation and route choice preferences.

As noted above and discussed in Chapters 8 and 10 in detail, with smart card data it is also possible to extract supply side data, such as the dwell time distribution at a bus stop. Therefore, it becomes possible to analyse mechanisms of "bus bunching" in detail. Most bus bunching studies focus on methods reducing its effect, but, to our knowledge, there are only few studies aiming to explain the causes of bus bunching with practical data so far an exception is Arrigada et al. (2015). With smart card data, it becomes possible to estimate the number of boarding passengers so that one can analyse the relationship between the demand and the supply service reliability.

## 5. BOOK OVERVIEW, WHAT IS MISSING AND CONCLUSION

The idea for this book was initiated following presentations given during the 1st International Workshop on Utilizing Transit Smart Card Data for Service Planning. This event was held in Gifu city, Japan on 2nd-3rd July, 2014. The objectives of this workshop were;

1. to create a network of researchers analyzing smart card data for further continuous exchange,
2. to exchange experience on how public transport smart card data can be best analysed with the final goal to establish some "best practice" guidelines,
3. to better understand that how far the data have been already utilized in practice, and
4. to include public transport operators in the ongoing (academic) discussion to better understand how they see the need and potential for smart card data analysis.

The workshop was attended by 45 participants from all over the world and included 23 presentations related to smart card data analysis. At the workshop, the participants agreed that the importance and potentials of smart card data deserve a book publication on how to use smart card data for public transport planning and evaluation.

The book is split into three sections. The first section aims to give an overview on estimating the different behavioural dimensions that can be analysed with smart card data. Firstly, Hickman discusses the various approaches to get transit origin-destination matrices from smart card data, considering that the smart card records often do not include both boarding and alighting record. Chapter 3 by Ali and Lee thereby discusses approaches to further infer activity types of passengers. Chapter 4 by Raveau concludes Part 1 by discussing challenges and possibilities to estimate route choice of passengers from smart card data. Taken together, if ODs, activities and routes of passengers can be estimated, then the analyst has a fairly complete overview on the travel patterns of passengers in the network and further indices such as network travel time can be extracted.

Part 2 discusses further analyses possibilities if smart card data are combined with other data sources. Chapter 5 by Kusakabe et al. discusses how smart card data could be fused with personal trip data, one of the challenges discussed afore. This is in fact also the bases for activity estimation of passengers, so that there is some overlap to Chapter 3.

Chapters 6 and 7 both offer a different perspective on the usage of smart card data in combination with survey data. For both the chapters the key is that the smart card usage and the survey response can be linked. In Chapter 6 by Brakewood and Watkins this is the key to estimate changes in the transit usage after installing real-time information. In Chapter 7 by Nakamura et al. sensitivities to the transit usage in response to a change in the loyalty-point scheme are analysed through a stated preference survey.

Chapter 8 by Fourie et al. combines smart card data with transit feed and other data to use these as input for activity based simulation. It further assesses the supply characteristics from smart card data and provides a powerful example on how smart card data can be used for a large-scale citywide simulation of the public transportation network. The chapter can hence be seen as a transition to Part 3 of the book which discusses how smart card data can be used to evaluate the transport network quality.

Chapters 9 and 10 directly focus on evaluation measures. The chapter by Munizaga et al. particularly discusses service indicators of interest for citywide transport planning. These are, for example, fairness in travel time distribution to the city centre from different parts of the city. Trepanier and Morency instead focus on evaluation measures of interest directly for service operators, such as service reliability, distance operated but also fare evasion.

Chapters 11 and 12 both discuss specific applications, though of very different kind. The chapter by van Oort et al. discusses ridership predictions in The Hague considering demand elasticity and potential changes in the service characteristics. Ishigami et al. discuss in Chapter 12 a basic application of smart card data where ridership information obtained from smart card data is used in combination with probe car data to assess the need to improve the environment of specific bus stops. Finally, Wilson and Hemily conclude this book in Chapter 13 by broadly looking at automatic data collection systems and pointing out further research areas.

The authors want to conclude this introduction by stressing that this book clearly does not offer a complete overview of all the existing smart card data research and some areas are missing. An important area that is not sufficiently covered in this book is discussions related to "within dynamics" as well as "day-to-day dynamics". To give an example of the former, smart card data can be used to discuss the network demand dynamics following an incident on one of the lines. An example for the latter might be Kurauchi et al. (2014) who discuss variation in the bus line choice of commuters with London Oyster data. Thus, these are some examples where further research is needed. In conclusion, since the discussion paper of Bagchi and White (2005) titled "The potential of public transport smart card data" some of these potentials have indeed been realized by now and the field has significantly advanced. However, to completely overcome some of the challenges that come with smart card data and to use their full potential will need further efforts. It is hoped that this book provides some overview of the state-of-the-art and will motivate scholars as well as practitioners to further advance the field.

## REFERENCES

Arriagada, J., Gschwender, J. and Munizaga, M. 2015. Modelling bus bunching using massive GPS and AFC data. *Proceedings of Thredbo 14*, Santiago de Chile, September.

Bagchi, M. and White, P.R. 2005. The potential of public transport smart card data, *Transport Policy*, 12 (5), September , pp. 464-474.

Harding. 2013. Big data econometrics. *Statistical Significance in Big Data.* Available from <https://bigdataeconometrics.wordpress.com/2013/12/28/statistical-significance-in-big-data/>. Accessed January, 2016.

Imai, R., Iboshi, Y., Nakamura, T., Morio, J., Makimura, K. and Hamada, S. 2012. Consideration on practical use of trail data acquired by smart card of transportation. *Proceedings of Infrastructure Planning,* Vol. 45, CD-ROM.

Jang, W. 2010. Travel time and transfer analysis using transit smart card data. *Transportation Research Record: Journal of the Transportation Research Board,* No. 2144, pp. 142-149.

Korea Smart Card. 2016. Homepage <http://eng.koreasmartcard.com/>. Accessed January, 2016.

Kurauchi, F., Schmöcker, J.-D., Shimamoto, H. and Hassan, S.M. 2014. Variability of commuters' bus line choice: An analysis of oyster card data. *Public Transport,* 6, pp. 21-34.

Pelletier, M., Trepanier, M. and Morency, C. 2011. "Smart Card Data Use in Public Transit: A Literature Review", *Transportation Research Part C,* 19, pp. 557-568.

## AUTHOR BIOGRAPHY

**Jan-Dirk Schmöcker** is an Associate Professor in the Graduate School of Engineering at Kyoto University. Jan-Dirk's research interests include a wide range of public transport issues, including modelling of network flows as well as data driven analysis of passengers' travel behaviour. He has published work related to analysis of London's Oyster card data and has been involved in studies using smart card data from Japan. Together with Fumitaka Kurauchi he initiated the 1st workshop on smart card data for transit planning in Gifu, Japan.

**Fumitaka Kurauchi** is a Professor in the Faculty of Engineering at Gifu University. His research interests include travel behaviour under provision of dynamic traffic information, modelling of transit network flows and network reliability analysis. He is a member of International Scientific Committee of Conference on Advanced Systems in Public Transport (CASPT). He published several analyses using smart card data such as London' Oyster card data. Together with Jan-Dirk Schmöcker he initiated the 1st workshop on smart card data for transit planning in Gifu, Japan.

**Hiroshi Shimamoto** is an Associate Professor in the Faculty of Engineering at University of Miyazaki. His research interests include passengers' travel behaviour analysis and road network analysis as well as public transportation network analysis. Among others, he is interested in network design issues and fare policy and how effects of potential service quality changes could be estimated with smart card data.

# PART 1

# Estimating Passenger Behavior

Chapter 2

# Transit Origin-Destination Estimation

M. Hickman[1,*]

**ABSTRACT**

Smart card transactions represent a passively collected source of information on passenger travel. With geographic coordinates and time stamps for these transactions, it is possible to infer the passenger's origin and destination of a journey. In cases where only one transaction takes place at the origin stop during a journey or trip leg (a so-called "tap-on"), an alighting location must be inferred. This chapter reviews the common methods and assumptions guiding inference of destinations. To supplement this review, it considers methods that convert the origins and destinations from smart card transactions into estimates of origin-destination flows (O-D matrices). Such estimates may be complicated by the interpretation of the smart card data, particularly with respect to activities that might occur at transfer locations. Finally, this chapter explores other methods employed to look at patterns in O-D journeys and in passenger tours throughout a day. Several avenues for continuing research in these areas are highlighted.

## 1. INTRODUCTION

Many cities and regions around the world have adopted smart card technology for fare payment, providing financial benefits to the public transport operator and convenience to the passenger. The smart card transactions are electronically recorded, commonly providing data about the time of transaction, identity of the card (e.g., a serial number and the card type), the fare charged and location of the card reader; e.g., at a rail or bus station, or on board a bus or light rail vehicle. In many cases, other

[1] School of Civil Engineering, University of Queensland, AEB (Bldg. 49) Room 551, St. Lucia, Queensland 4072. Email: m.hickman1@uq.edu.au

* Corresponding author

data might be collected, such as a vehicle identifier, the route number, the travel direction and whether the transaction was an originating trip leg or a transfer from an earlier trip leg. For large transit networks, a single day's operation may yield hundreds of thousands or millions of smart card transactions.

While these data are primarily collected to manage fare collection, the availability of these data is certainly very attractive to public transport planners: the data are passively collected, without requiring more expenditure, and in many cases represents a large or nearly complete sample of journeys or trip legs made by public transport.[1] Historically, ridership data were often collected manually, infrequently, at a huge cost and of varying quality. Hence, the change from a relatively "data-poor" environment to a very "data-rich" environment creates many new opportunities to analyse transit ridership patterns and to improve public transport service planning (Bagchi and White 2005; Pelletier et al. 2011). Perhaps understandably, the data come with errors, inconsistencies and missing values that are in part unique to smart card data, but which can be managed through various techniques (Utsunomiya et al. 2006; Zhao et al. 2007; Robinson et al. 2014; Yang et al. 2015).

This chapter specifically addresses how to find transit passenger origins and destinations, as well as possible journey patterns, in one or more days of smart card transactions. Trip legs, a complete journey, the combination of journeys in a tour and related features of repeated journeys represent very practical measures of transit ridership. These data can offer a useful snapshot of individual passenger (disaggregated) travel patterns, may show changes in travel patterns over time, can show changes in demand in response to service or fare changes or changes in exogenous variables and may help planners to forecast future changes in ridership and passenger travel patterns for changes in service.

While data from smart cards can help to show passenger travel, the primary function of the smart card is to pay a fare. Hence, the design of a smart card system is to facilitate fare payment using local fare policies and structures. Conversely, the smart card system and its data usually do not directly serve the data needs of transit planners. This chapter explores the features common to smart card data and possible methods of improving their use to describe passenger origins, destinations, time of travel and travel patterns. Section 2 describes the basic features of smart card data that assist with its interpretation of passenger trip legs and journeys. Subsequently, Section 3 includes a formal review and discussion of destination inference and its assumptions and violations. Section 4 highlights work in origin-destination (O-D) matrix estimation and methods

[1] The word *journey* describes the movement of a person from their origin to their destination. Within a journey, a passenger will have one or more *trip legs*: each trip leg makes up the passenger movement associated with a single vehicle. Thus, a transfer to a second vehicle begins a second trip leg.

to infer routes, transfers and intermediate activities. Finally, Section 5 discusses recent data mining and analytic methods to explore passenger trip purpose as well as journey and tour patterns. Brief comments on future areas of research are given in Section 6.

## 2. GENERAL PRINCIPLES

The use of a smart card involves tapping, swiping or waving the card on or over a reader either at the stop/station or on boarding the vehicle. A flat fare policy and some zonal fare policies only require that a passenger taps once, either before boarding at a station, or while boarding the vehicle. In these cases, it records only a single transaction (a "tap-on"). More complicated fare policies based on distance or zones usually require that the passengers tap-on and tap-off with the smart card.

Thus, interpretation of a tap-on and/or tap-off transaction depends in part on the fare policies and transfer policies within the transit network. In the simplest case, a single tap-on or a joint tap-on and tap-off indicate a single trip leg. In "closed" transit networks where no tap is necessary at an interchange (e.g., in a rail network), a single tap-on is all that is available to interpret the full passenger journey. In "open" networks, passengers must tap-on for each trip leg with a separate transaction record for each trip leg in a journey.

To understand the trajectory of a passenger, it is often useful to match the time and location of the passenger tap with the time and location of a vehicle. This matching might be done with some additional processing of automatic vehicle location (AVL) data, which records the location and time-stamp of vehicle movements at stops and along a given route (e.g., Barry et al. 2002; Zhao 2004; Zhao et al. 2007; among many others). However, this matching also requires a common time and spatial reference between the smart card and AVL systems. In the absence of AVL data, explicit matching of passenger movements to scheduled bus and train movements might be difficult. A *de facto* schedule data format, such as Google's General Transit Feed Specification (GTFS 2015), might be used. However, if schedule adherence is low or headway variability is high matching of a vehicle location and time to a passenger's tap requires extra effort.

Some common assumptions are often implicit in smart card analysis. First, a smart card ID is usually presumed to represent a single passenger ("nontransferable"), allowing interpretation of smart card transactions as the movements of one passenger. However, if there is sharing of the card, the movements cannot be easily reconciled to a single person. Second, the fare payment and transfer policies may themselves influence passenger behaviour. Examples of policies that could alter behaviour include transfer discounts, free trips after a maximum daily fare or maximum daily number of journeys, free trips after a certain daily (weekly, monthly) maximum, or some daily (weekly, monthly) maximum fare payment. In these situations, passengers might be willing to game the system in order to achieve fare savings. This, in turn, may lead to differing interpretation of passenger behaviour, if that behaviour is strongly affected by the fare policies.

## 3. INFERENCE OF DESTINATIONS

The tap-on of a smart card is usually sufficient to identify the origin and the starting time of the trip leg. If the destination of a trip leg (alighting location)[2] and time of arrival is desired, one needs either (1) an additional tap-off from the smart card, or (2) a means to infer this destination. Due to the prevalence of single tap-on systems, many researchers have investigated the problem of inferring destinations.

### 3.1 Tour ("Trip Chain") Assumptions

The most common technique of inferring the destination and time of arrival uses the notion that a "tour" or "trip chain", describing the chain of trip legs that a person will make within a single day. The chain assumes that the destination of one trip leg is proximate to the origin of the next trip leg and that the destination of the last trip leg in the chain is proximate to the origin of the first trip leg. The chaining assumption also infers that no journey during the tour is done by a different (non-walking) mode. Logically, a tour requires that the person will travel at least two trip legs.

An example for trip chain is shown in Figure 1. A passenger leaves home for the first destination and as a part of that journey it is necessary for him/her to make a transfer. Transactions (tap-ons) are recorded when boarding at the origin and when boarding at the transfer; however, the locations of alighting on the first and second trip legs are not known. The passenger then makes a second journey and third journey, to return to the tour origin. As noted in the figure, the smart card transactions give the origins and time of departure of each trip leg.

The problem, then, is to infer the transfer or destination locations. As one technique, one may choose the closest stop on the previous route, nearest to the next transaction. In Figure 1, the alighting point on the first bus might be inferred as the stop on that route nearest to the second transaction. Similarly, the alighting point from the second bus could be inferred as the stop on the second route nearest to the third transaction site. If the passenger's alighting time is also desired, a common approach is to estimate the time the bus arrives at that location; this time could be determined either from AVL records or from the scheduled time on that bus route.

[2] In most literature, this is called a "destination" and the process "destination inference". However, in the scope of this chapter, what is meant is an "alighting location", as the passenger may only be making a transfer. In keeping with this literature, it uses the word "destination", but with this caveat.

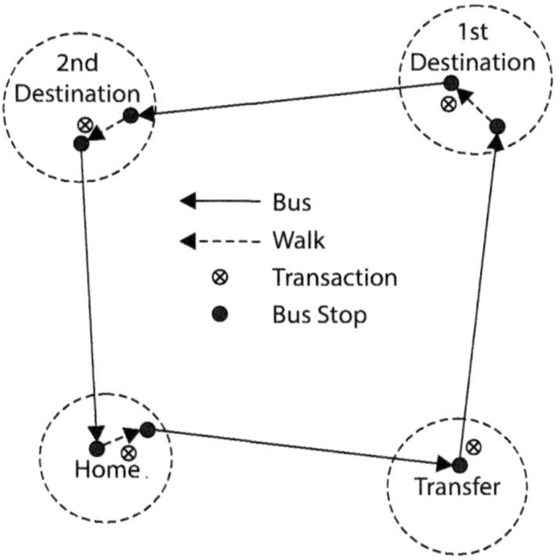


**Fig. 1.** Trip leg and journey chaining model

Common assumptions needed within the trip chaining model include:

1. The destination of the last trip leg in a tour is identical to the origin of the first trip leg in the tour.
2. Passengers will generally take the most direct walking paths between services, as measured by time, by distance, or some generalized time or cost.
3. Passengers will take the next service available after arriving at a station/stop.

One may use assumptions 1-3 to infer the most likely stop at the end of a trip leg and to compute the time spent transferring between two services. If there is no time-consuming activity or long walk required during the interchange, the assumption is that the passenger will continue their journey directly by taking the first subsequent boarding opportunity.

## 3.2 Inference Methods

Most methods to infer destinations build from the simple algorithm suggested before. For each trip leg where the alighting location is unknown, infer the alighting location as the nearest stop on the route that is closest, in distance, to the next transaction. If there is no further transaction for the day, infer the alighting location as the stop on the route that is closest to the first transaction of the trip chain. Generally, one might assume certain maximum distances might apply, to avoid violating assumptions 1 and 2 above and to identify if the trip chain is interrupted by longer, non-walking trips. The algorithm fails to produce an alighting stop if these maximum distances are exceeded, or if the passenger only has a single trip leg or single journey on the given day.

As an example, Barry et al. (2002) used this algorithm to infer destinations in the New York City subway. To certify the trip chain assumption, they employed a sample of 100 passengers who made only 2 journeys and 150 passengers who made chains of 3 or more journeys in a single day. In both samples, 90% of destinations could be successfully inferred. Then, using the subway fare card data of a single day, destinations could be successfully inferred for 83% of subway fare card transactions, with the lower fraction attributed to fare card errors or to those cards observed for only a single journey. The O-D patterns of fare card users were then expanded to include all subway passengers (including the 22% without fare cards), with the assumption that non-fare card passengers share the same O-D patterns with fare card users. Station-specific boarding and alighting counts and passenger counts across selected cordons were used to show the validity of O-D flows.

Two improvements to this destination inference algorithm were suggested by Trépanier et al. (2007). First, in cases where multiple days of smart card data are available, the last alighting location on a given day is given as (1) the initial boarding location of the tour, if the route is identical to the first route taken; or, (2) the initial boarding location of the first journey on the subsequent day. Second, for those trip legs where an alighting location cannot be inferred otherwise, the destination might be inferred as an alighting point for the same passenger, if he/she historically has used the same route and boarding stop. With these improvements, about 66% of alighting locations were successfully inferred, taking into account erroneous smart card data (21%) and trip legs with no successful inference (13%). Rates of inference were higher for more heavily used routes, for frequent travellers and for the morning peak period, when compared with infrequent travellers or travel in the off-peak, late evening and weekend periods. These two improvements were enhanced by the work of Ma and Wang (2014), who developed a Bayesian decision tree to classify historical origins and destinations. This decision tree then creates other probable inferences for a trip leg destination when other trip-chaining criteria are not satisfied.

For a multi-modal system, Zhao (2004) and Zhao et al. (2007) added an additional rule to the basic algorithm: the symmetry in routes in a daily tour (e.g., mirrored rail-rail or rail-bus route sequences) could infer alighting locations, if these were not otherwise identified. For a week of fare card data, about 71% of alighting locations could be successfully inferred. Farzin (2008) and Wang et al. (2011) used a similar approach to perform destination inference.

A different passenger aim to infer alighting locations in bus-to-bus transfers was introduced by Munizaga and Palma (2012). Their approach minimizes the total time, defined as the time onboard plus time spent walking from an alighting location to the next boarding site, to infer the alighting stop. These objectives have an advantage of finding locations that minimise the passenger transfer time.

The common assumptions for destination inference were tested empirically using the data from some cities that have tag-on, tag-off data. Notably, Alsger et al. (2015) used data from Brisbane, Australia to explore the largest walking distance, the greatest transfer time and the destination for the last journey of the day. First, they observed that, for transfer time of up to 90 minutes, the distance from an alighting stop to the next boarding stop rarely exceeds 800 m. They then concluded that 800 m is a reasonable maximum for identifying potential transfers. They also observed that transfer walking distances are relatively short, with about 80% of walk time being less than 5 minutes and over 90% of walking time being less than 10 minutes. Second, they note that the total number of journeys with an inferred transfer ranges from 15% to 20% of journeys, as the assumed transfer time threshold rises from 15 to 45 minutes. Only a very slight increase in the percentage of journeys with a transfer occurs when the allowable transfer time value is increased up to 90 minutes. The conclusion, supplemented by statistical evaluation of matrix similarity, is that the origin-destination matrix is not affected significantly by the assumed transfer time. Finally, they observed that 82% of tours returned to the same stop at the end of the day, while 90% were within 400 m of their tour origin and 95% were within 800 m of their tour origin from the same day.

In a separate study, He et al. (2015) investigated destination inference quality, using tag-on, tag-off data from Brisbane as the ground truth. Their method, based on Trépanier et al. (2007), inferred the correct destination for 66% of trip legs. However, their analysis showed that, for a given distance threshold, there are a number of potential stops that might serve as reasonable destinations (e.g., among a high density of stops in the central business district). As a result, correct destinations were identified, if allowing all stops within a given distance, rather than using the minimum-distance stop. For example, by including possible "near misses" at 400 m, successful inference of the true alighting stop improves to 79%. Improvements in inference by allowing "near misses" are largest for trip legs on weekdays as compared to weekends and for peak periods (5-8 am, 4-7 pm weekdays) as compared to off-peak periods. Nonetheless, the accuracy of the destination inference is relatively insensitive to the real value of the distance threshold for "near misses".

## 3.3 Transfer vs. Activity Inference

One challenge in inferring journey destinations is that the passenger may take part in short-duration, location-specific activities that are not easily discriminated from a transfer, especially if the transfer policies are generous. For example, if the fare policy allows transfers up to 60 minutes, passengers may conduct a short activity and return to their origin, but this is recorded as a transfer. Hence, differentiating transfers from a location-specific activity is not usually revealed in the smart card data. Some activities might be merely incidental to the transfer (e.g., buying a newspaper or a beverage), while in other cases, a location-specific activity of the passenger (e.g., shopping, a meeting with a friend) occurs. Separating transfers from true location-specific activities is important in capturing true passenger origins and destinations.

Initial research used a simple time threshold to distinguish transfers from activities. Hofmann and O'Mahoney (2005) used a 90 minute interval, while Bagchi and White (2005) used a 30 minute interval, between separate boarding transactions (from tap-on to tap-on). Both the teams suggested that this interval be conditioned on the size of the city, with larger cities allowing greater time between boardings. In another investigation, Barry et al. (2009), used a 18 minute maximum gap from alighting to next boarding to infer a transfer, while Munizaga and Palma (2012) used a 30 minute gap. Jang (2010) showed that transfer times were less than 10 minutes for 80% of journeys involving a transfer in Seoul.

A proposal was given by Chu and Chapleau (2008) and Chu (2010) for a more rigorous accounting of the time between alighting and a subsequent boarding. In their study, they calculated the time of alighting and added the estimated walk time to reach the transfer stop, with a 5 minute buffer added for any uncertainty in the connection. If the passenger is observed, to take the next available vehicle on the connecting route, it infers a transfer; if not, the passenger is inferred to have conducted an intermediate activity. This more careful consideration of the timing of transfers results in a decrease in the estimate of multi-leg journeys (almost 40% in this case), compared with simply using a maximum transfer distance to find transfers. In Nassir et al. (2011), similar rigorous accounting was used to infer destinations and to identify incidental and destination-specific activities.

To account for incidental activities during a transfer, Seaborn et al. (2009) consider developing separate thresholds to find maximum possible time for subway-to-bus, bus-to-subway and bus-to-bus transfers. Notably, their analysis suggests that the nearest transfer is not always the one taken if the incidental activity takes a short period or involves a longer walk. A systematic study of these transfers in London resulted in recommendations for thresholds of: (1) 15-25 minutes for subway-to-bus transfers (subway station tap-off to bus tap-on); (2) 30-50 minutes for bus-to-subway transfers (tap-on upon bus boarding to tap-on at a station); and (3) 40-60 minutes for bus-to-bus transfers (tap-on upon one bus to tap-on upon the next bus).

Two fairly intuitive criteria described in Devillaine et al. (2012) work in conjunction with a 30 minute transfer time threshold: (1) the person exits and then re-enters a rail system; or, (2) the person travels again on the same route in the bus network. In these cases, intuition suggests an activity was conducted, regardless of the duration.

A major study presented by Gordon (2012) and Gordon et al. (2013) suggests a comprehensive set of rules to differentiate transfers from short activities, using smart card data from London. It assumes a transfer, unless one of the following is true:

- The trip leg is the last one of the day.
- The inferred alighting stop is more than 750 m from the next boarding stop.
- The passenger boards the same route from which they most recently alighted.
- The resulting journey destination is less than 400 m from the origin of journey.
- The transfer time exceeds the maximum time, including the walking time (of at least 5 minutes) to the next boarding stop, plus the minimum of a 45 minute waiting time or the time of the next scheduled arrival of a bus at the boarding stop.
- The circuity of the trip, measured by the real distance travelled divided by the straight-line distance, exceeds some threshold (e.g., 1.7).

  The use of these criteria in a London Oyster card case study resulted in 22% of connections being classified as transfers, 69% classified as activities and 9% as unknown. Such a characterization of activities results in a set of passenger origins and destinations.

  Nassir et al. (2015) build upon the work of Gordon et al. (2013) to evaluate several criteria to infer a transfer or an activity. It concludes a transfer unless:

- The passenger boards the same route from which they most recently alighted.
- The resulting journey destination is less than 400 m from the journey origin.
- The transfer time (gap) exceeds a minimum time (e.g., 20 minutes).
- The ratio of the gap to the total travel time exceeds some ratio (e.g., 0.4), suggesting that the intervening time consumed a substantial fraction of the total travel time.
- The circuity of the trip, measured by the real distance travelled divided by the straight-line distance, exceeds some threshold (e.g., 1.7).
- The difference between the observed travel time and the least travel time (so-called "off-optimality") for the origin-destination pair at the given time of day exceeds some minimum time (e.g., 20 minutes).
- The ratio of the off-optimality to the total travel time exceeds some threshold (e.g., 0.5).

These criteria are developed and empirically calibrated by comparing transfers to and from the same route, which are interpreted as a result of intervening activities, with transfers among different routes. These differences are plotted in the space of gaps, travel times and off-optimality, to derive the specific values used in a case study of Brisbane. Their results suggest that, among almost 2 million sequences from March 2013 with two or more trip legs separated by less than 60 minutes, about 414 thousand (21%) might be inferred as including a location-specific activity.

## 4. O-D MATRIX METHODS

A rather simple interpretation of the origins and destinations (O-Ds) emerging from smart card data is that the data can simply be fed directly into an origin-destination matrix (Buneman 1984). Using a given seed matrix, or the smart card data itself as a seed matrix, common matrix expansion methods (iterative proportional fitting, the Furness method, maximum likelihood estimation, etc.) might be exploited to estimate the true O-D matrix (Cui 2006, Lianfu et al. 2007; Park et al. 2008; Li et al. 2011; Zhao 2004; Zhao et al. 2007). In some cases, other information sources can supplement these estimates; for example, Frumin (2010) uses estimates of train loads from weight sensors to help in the passenger assignment and O-D estimates on a rail line. Because of the multitude of available paths, Gordon et al. (2013) use expansion methods based on individual O-D paths, rather than the aggregate O-D flows.

There are many considerations, however, that may affect how useful such a matrix might be, for the purpose of estimating the true origin-destination flows in the public transport system (Gordillo 2006; Chan 2007). Those challenges include:

***The ratio of passenger journeys using smart cards***, as compared to all passenger journeys. In some systems, the percentage of passengers using the smart card could be high (e.g., 85-90%), representing a very large majority of trips. However, one must be careful even at these high percentages for possible differences in travel behaviour among smart card users and non-users. If there are major differences in the time of travel, the origin and destination locations, the types of daily tours, the frequency of travel, fare evasion, or other travel behaviour, simple factoring of the smart card O-D flows might be biased and misleading (Gordillo 2006; Munizaga and Palma 2012).

***Self-selection bias*** among those who use the smart card. As one example, one might expect that passengers who use the public transport system often, or who otherwise might not pay fares by other means (e.g., cash, weekly or monthly passes, or discounts over cash) might be more likely to use a smart card. In this case, this population may have different travel characteristics than more infrequent users or pass-holders. As a second example, the smart card might target certain groups: primary and secondary school students, employees of certain businesses or government, pensioners/retirees, university students and staff, etc. In these cases, one expects there might be clear differences in the trip-making behaviour of these groups, compared with the universe of public transit users (Lee and Hickman 2011, 2013, 2014).

***Temporal and behavioural differences in the meaning of the tag-on.*** With time stamps at the tag-on, it is possible to generate time-dependent O-D matrices, under the assumption that passenger flows are reasonably uniform over the time period of interest (for example, see Ji (2011) and Ji et al. (2011) for estimating these time intervals). However, mixing of data from tag-on on-board with that off-board could be slightly inconsistent. A tag-on at a stop/station occurs when the passenger arrives, compared to a tag-on while boarding a vehicle. As a result, for time-dependent O-D matrices that combine both off-board and on-board transactions, it is important to consider a consistent point of time from the passengers' perspective; e.g., one may use an inferred boarding time, for modes or services where the tap-on occurs at a stop/station.

***Mapping O-D flows from stops to flows from traffic analysis zones.*** As most transportation planning models are based on the geographic unit of the Traffic Analysis Zone (TAZ), it is not easy to map O-D flows based on transit stops to the more general geography of TAZs. For example, stops might be located along roadways along the border of a TAZ, requiring a stop-to-TAZ (many-to-one) assignment. Instead of assigning all flow to the nearest TAZ, others have sought to capture the catchment areas of a stop more carefully. Most recently, the work of Tamblay et al. (2015) provides a fractional assignment of stops to TAZs using a logit model, built upon passenger walk access data from zonal data, land use data and a passenger access survey.

The challenge in the first two cases is to find information on the sources of bias and to use this information to expand the O-D flow estimates. In many cases, such additional information will rely on independent household travel surveys, passenger on-board surveys, or other observational studies that capture different passenger types. Ideally, existing household travel surveys and passenger on-board surveys would also collect information on the serial number (ID) of any smart card used, to validate public transport use for smart card users and to correct for these possible biases among non-users (Chapleau et al. 2008; Munizaga et al. 2014; Kusakabe and Asakura 2014).

## 5. JOURNEY AND TOUR PATTERN ANALYSIS

There is a growing literature which seeks to describe travel using not only time-dependent O-D matrices, but also to capture disaggregated travel patterns across many days. Patterns such as the frequency of travel, the timing of travel, journey origins and destinations and passenger trip chains could be used to classify passenger behaviours, to measure the variability of those behaviours and to give other meaningful aggregations of passenger movements. This section examines common challenges in these data analyses.

## 5.1 Identification of Routes from Smart Card Data

One of the common difficulties is the inference of the passenger's routes, one for each trip leg, when this information is not included among the data collected in the fare system. For example, when a tap-on occurs at a station, there could be some uncertainty about when the passenger actually boarded a vehicle and boarded which route. If the route and direction are not identified, these characteristics then might be inferred.

Methods to infer route and direction commonly rely on observations of the passenger's time of departure from the origin, time of arrival at the destination and the resulting travel time; also, methods based on travel distance and/or transfers could be employed (Reddy et al. 2009). These passenger-specific times are observed from the smart card data; what is commonly missing is the assignment to a specific route or scheduled vehicle run. One common method to achieve this assignment is to generate a set of feasible paths from the origin to the destination, using a time-dependent shortest path algorithm. The time-dependent travel times in this algorithm track individual train or bus movements in the network and could be taken from the published timetable or available AVL data. Various methods could be employed to select the most likely combination of routes and vehicles for the passenger, given the passenger's observed travel time characteristics.

In this area, examples of research using a deterministic, rule-based method include Kusakabe et al. (2010); Asakura et al. (2012); Zhou and Xu (2012); Sun et al. (2012); Van der Hurk et al. (2015); Hong et al. (2015); and Sun and Schonfeld (2015). Extensions of these rule-based methods to examine passenger "strategies", where passengers may make boarding decisions based on the timing of vehicle arrivals to the origin (so-called "hyperpaths") are explored by Schmöcker et al. (2013) and Kurauchi et al. (2014).

In other cases, probabilistic considerations may dominate. Notably, in a study from London's underground network, Paul (2010) considered the means of estimating passenger routes and trains. Use of smart card data to estimate the travel time from one station to another was matched with the trajectory data of the trains. The path was inferred from the possible train trajectories and the probability distribution of passenger walking times, explicitly considering platform access and egress times and transfer times within each station. Paul's work was extended to the Hong Kong MTR in the work by Zhu (2014). Alternately, advanced algorithms can simulate a passenger's path-specific travel time and explore the resulting O-D travel time distributions, to infer the most likely path of the passenger. Bayesian frameworks such as Markov Chain Monte Carlo (MCMC) simulation (Lee and Sohn 2015) or Metropolis-Hastings sampling (Sun et al. 2015) are data mining techniques that have been explored. In a different approach, the research by Fu et al. (2014) used Gaussian mixture models to explore passenger's travel time distributions in London's underground network, to find different routes used by passengers. For further discussion on route choice estimation with smart card data see Chapter 4 in this book.

## 5.2 Journey Pattern Analysis

There might be value in analysing similar travel patterns among groups of passengers, for the purpose of understanding existing and potential transit market segments and for generating possible information and service strategies for these markets. The use of smart card data for this task provides another level of disaggregation. This is an emerging area of research, using data mining and trajectory clustering techniques to illuminate important passenger behaviours.

At a basic level, statistical methods for the analysis of passenger travel patterns include frequency analysis, ANOVA and related spatial and temporal correlations among journeys (e.g., Nishiuchi et al. 2013 among many others). Visually, the work of Tao et al. (2014a, 2014b) explores the illustration of mapped passenger O-D flows using a so-called "flow co-map". Such co-maps are extensions of existing passenger flow diagrams, but in this case aggregation of each journey in time and space and various conditions (e.g., direction of travel, use of a busway) could be employed to illustrate specific types of passenger flows during different times of the day.

Using clustering methods, many researchers have sought to look at temporal and spatial travel patterns, usually by origin and destination and by time of the day. K-means clustering was used by Zhao et al. (2014) to identify the typical spatial and temporal travel patterns and to identify "anomalous" behaviour that does not easily fit existing clusters. Yuan et al. (2013) use Conditional Random Fields (CRF) to identify passenger journey chains from spatial, temporal and card transaction constraints. The goal in this work was to discover both passenger boarding and alighting locations as well as tour-based mobility and activity patterns. A Naïve Bayes classifier was used by Foell et al. (2013, 2015) to classify passenger trips based on the day of week, time of day and frequency of travel. An extension of this model to predict passenger boarding sites is described in Foell et al. (2014). Kieu et al. (2015) extended the traditional DBSCAN algorithm to consider the density of bus stops in the vicinity of a location to infer passenger travel patterns through tours. This algorithm takes as input the location and time stamps of journeys or tours and allows the user to specify various tolerances in space and time. From this information, the algorithm then clusters passenger journeys or tour patterns into common or shared patterns.

A separate line of investigation has looked at identifying travel patterns of specific passenger market segments; this could be important in public transport marketing, information strategies and in determining passenger response to service changes. K-means clustering was used by Agard et al. (2006) and Morency et al. (2007) to investigate the temporal and spatial variability of travellers who use various types of smart card. El Mahrsi et al. (2014) used K-means clustering to group passengers into types based on their temporal travel characteristics (hour-of-day and day-of-week). With a similar objective, Kieu et al. (2014) used DBSCAN to segment public transit passengers based on their day-to-day travel patterns, both in space and time. Similarly, Costa et al. (2015), compared three different machine learning techniques (decision trees using J48, Naïve Bayes and Top-K algorithm) to classify passenger travel patterns into four groups, based on the level of spatial and temporal regularity of their journey patterns. Spatial and temporal clustering of passenger travel patterns has also been explored in Lathia et al. (2010, 2013) using a dendrogram as a form of agglomerative hierarchical clustering. In a contrasting approach, Ma et al. (2013), used DBSCAN to cluster an individual traveller's journeys, based on the spatial and temporal dimensions of their journeys and tours. These passenger-specific clusters in turn are clustered with other travellers' travel patterns using the K-means++ algorithm. The authors also explore the use of rough-set theory to create a rule-based classifier from the K-means++ results. The rough-set theory-based classifier is used to identify similar journey clusters for passenger journeys with only a tap-on.

## 5.3 Activity Inference and Analysis

While the data from smart cards does not include any information on the activities conducted by passengers during their daily tours, some have explored extensions of the journey patterns, trip chains and land use data at journey destinations to infer possible passenger activities. As with journey pattern analysis, this allows planners to understand the existing and potential passenger markets and potential strategies to attract more passengers to public transit. Knowing the activity type (mandatory vs discretionary) also allows a deeper understanding of possible passenger responses to transit service changes.

One direct form of analysis is to look at repeated destinations that passengers visit over time. As one example, the work of Chu and Chapleau (2008) was extended in Chu and Chapleau (2010) to identify trip "anchors", representing frequently used stops in a small vicinity of a given destination (e.g., within a 500 m radius). These anchors might be associated with home, work or school locations, depending on the local land use at that destination. Extensions to model passenger activity patterns, using decision trees with the C4.5 algorithm, were also explored in this research.

Other investigations have explored other travel patterns shown in the smart card data to derive trip purposes. Bouman et al. (2013, 2015) generate a set of rules to characterize passenger activity patterns using smart card data from the Netherlands. The critical data elements from the smart card transactions are the duration of the activity and the sequence of the activity in the overall trip chain (or the start and end time of the activity).

A major extension of this approach uses land use data at transit destinations and information on the smart card type to make further inferences about trip purpose. Devillaine et al. (2012), Lee et al. (2013), Lee and Hickman (2014) and Ali et al. (2015) each generates a set of rules to characterize the journey purpose (e.g., work, school, home, other), considering smart card transaction data that combines with GIS data on land use at destinations. The land use data is exploited to infer likely activities conducted near transit stops. The work of Munizaga et al. (2014) serves to validate these approaches, comparing the trip purpose inferred from the smart card with corresponding household survey data as well as other survey data.

Others have considered integrating household travel survey data, which provides trip purpose information, with the smart card data. Chakirov and Erath (2012) investigate the types of activities that could be identified from smart card data, particularly examining rule-based methods to classify work activities. These rules are not as effective, however, when compared with methods that integrate household travel survey data. Specifically, with the household survey data, the researchers generated logit models to predict work activities from the duration, start time and site of the activity, using detailed land use data at journey destinations. By applying these logit models to the smart card data, a larger percentage of work trips could be successfully inferred than using the simple rules.

Finally, Kuhlman (2015) uses smart card data to enrich local survey efforts to examine travel patterns and activities, comparing both journey-based and tour-based pattern analysis to infer passenger activities at destinations. The results suggest considerable benefits of expanding travel survey data with smart card data, to infer trip purpose, particularly for work journeys but also for "other" trip purposes; shopping and educational purposes were less accurately predicted. In addition, a tour-based approach, incorporating the full trip chain over the course of a day, has much better inference of trip purpose than a trip-based approach. This discussion is continued in Chapters 3 and 5.

## 6. AREAS FOR FUTURE RESEARCH

The use of smart card data to estimate passenger origin-destination flows, and associated extensions to tours, within-day travel and activities and travel patterns across days, represents a healthy area of research. The review in this chapter has illustrated a wide variety of research into methods of structured analysis of the smart card data and into applications for better transit planning.

While one might consider this area fairly mature, there are some areas where the value of the smart card data could be further exploited for similar applications. First, more effort is required to achieve integration of transit smart card data with household travel survey data and with passenger on-board survey data (Chapleau et al. 2008; Trépanier et al. 2009; Medina and Erath 2013; Munizaga et al. 2014; Kusakabe and Asakura 2014; Spurr et al. 2014; Kuhlman 2015). One might expect that future survey efforts might attempt to capture a passenger's smart card identifier and their permission to use smart card transactions as part of newer survey methods. In this way, rather than relying on travel surveys alone to capture passengers' travel patterns, panel data from smart cards could be used to monitor travel behaviour over longer periods of time, but importantly, the travel recorded by the smart cards could be connected to sociodemographic characteristics. Currently, there is no direct way of capturing this connection. Synthetic methods, where travel patterns from surveys are matched to specific fare card-revealed travel, may offer a close approximation to such direct integration.

Second, while there are clear methods to generate O-D matrices from smart card data, there remain some obvious questions that have not yet been answered. As noted earlier, the research community lacks good methods to find possible sample- or self-selection bias in generating O-D matrices from smart cards. It also does not have a good idea how much O-D matrices vary over time, such as on a day-to-day basis. Such information would be useful, from a service planning perspective, to understand the extent to which demand varies and the extent to which that variance is a function of demographic variables, service variables and perhaps other exogenous variables. Improvements in demand modelling may occur if such sources of variability could be identified.

Third, there are many opportunities to create a stronger connection between O-D estimates and their more practical use in travel demand modelling. The work of Tamblay et al. (2015) ties the smart card data to O-D estimates that could be directly connected to traditional traffic analysis zones (TAZs) used in strategic transport planning models. Also, Section 5.1 describes recent efforts to identify journey patterns, activity patterns and inferences of trip purpose; this work could be more directly related to travel demand management, transit demand forecasting and transit service design.

These are exciting times for researchers in these areas. Yet, at the risk of a gross generalization, there remains a pressing need to show the value of such analytic methods to improve the practice of service planning.

## REFERENCES

Agard, B., Morency, C. and Trépanier, M. 2006. Mining public transport user behaviour from smart card data. *In 12th IFAC Symposium on Information Control Problems in Manufacturing-INCOM*, pp. 17-19.

Ali, A., Kim, J. and Lee, S. 2015. Travel behaviour analysis using smart card data. *KSCE Journal of Civil Engineering*, pp. 1-8. doi:10.1007/s12205-015-1694-0.

Alsger, A.A., Mesbah, M., Ferreira, L. and Safi, H. 2015. Public transport origin-destination estimation using smart card fare data. *In Transportation Research Board 94th Annual Meeting* (No. 15-0801).

Asakura, Y., Iryo, T., Nakajima, Y. and Kusakabe, T. 2012. Estimation of behavioural change of railway passengers using smart card data. *Public Transport*, 4(1), pp. 1-16.

Bagchi, M. and White, P.R. 2005. The potential of public transport smart card data. *Transport Policy*, 12(5), pp. 464-474.

Barry, J., Freimer, R. and Slavin, H. 2009. Use of entry-only automatic fare collection data to estimate linked transit trips in New York City. *Transportation Research Record: Journal of the Transportation Research Board*, (2112), pp. 53-61.

Barry, J., Newhouser, R., Rahbee, A. and Sayeda, S. 2002. Origin and destination estimation in New York City with automated fare system data. *Transportation Research Record: Journal of the Transportation Research Board*, (1817), pp. 183-187.

Bouman, P., Van der Hurk, E., Kroon, L., Li, T. and Vervest, P. 2013. Detecting activity patterns from smart card data. *In BNAIC 2013: Proceedings of the 25th Benelux Conference on Artificial Intelligence, Delft, the Netherlands, November* 7-8, 2013. Delft University of Technology (TU Delft); under the auspices of the Benelux Association for Artificial Intelligence (BNVKI) and the Dutch Research School for Information and Knowledge Systems (SIKS).

Bouman, P., Van der Hurk, E., Kroon, L., Li, T. and Vervest, P. 2013. Detecting time interval patterns from smart card data. Available at http://rstrail.nl/new/wp-content/uploads/2015/02/bouman_paul.pdf.

Buneman, K. 1984. Automated and passenger-based transit performance measures. *Transportation Research Record: Journal of the Transportation Research Board*, (992), pp. 23-28.

Chakirov, A. and Erath, A. 2012. Activity identification and primary location modelling based on smart card payment data for public transport. Eidgenössische Technische Hochschule Zürich, IVT, Institute for Transport Planning and Systems. http://dx.doi.org/10.3929/ethz-a-007328823.

Chan, J. 2007. Rail transit OD matrix estimation and journey time reliability metrics using automated fare data (*MS Thesis, Massachusetts Institute of Technology*).

Chapleau, R., Trépanier, M. and Chu, K.K. 2008. The ultimate survey for transit planning: Complete information with smart card data and GIS. *In 8th International Conference on Survey in Transport, Lac d'Annecy, France.*

Chu, K.K.A. 2010. Leveraging data from a smart card automatic fare collection system for public transit planning. (*Doctoral dissertation, École Polytechnique de Montréal*).

Chu, K.K.A. and Chapleau, R. 2008. Enriching archived smart card transaction data for transit demand modelling. *Transportation Research Record: Journal of the Transportation Research Board*, (2063), pp. 63-72.

Chu, K.K.A. and Chapleau, R. 2010. Augmenting transit trip characterization and travel behaviour comprehension: Multiday location-stamped smart card transactions. *Transportation Research Record: Journal of the Transportation Research Board*, (2183), pp. 29-40.

Costa, V., Fontes, T., Costa, P.M. and Dias, T.G. 2015. Prediction of journey destination in urban public transport. *In Progress in Artificial Intelligence*, pp. 169-180. Springer International Publishing. doi:10.1007/978-3-319-23485-4_18.

Cui, A. (2006). Bus passenger origin-destination matrix estimation using automated data collection systems (*MS Thesis, Massachusetts Institute of Technology*).

Devillaine, F., Munizaga, M. and Trépanier, M. 2012. Detection of activities of public transport users by analyzing smart card data. *Transportation Research Record: Journal of the Transportation Research Board*, (2276), pp. 48-55.

El Mahrsi, M.K., Côme, E., Baro, J. and Oukhellou, L. 2014. Understanding passenger patterns in public transit through smart card and socioeconomic data: a case study in Rennes, France. *In ACM SIGKDD Workshop on Urban Computing* (p. 9).

Farzin, J. 2008. Constructing an automated bus origin-destination matrix using fare card and global positioning system data in Sao Paulo, Brazil. *Transportation Research Record: Journal of the Transportation Research Board*, (2072), pp. 30-37.

Foell, S., Kortuem, G., Rawassizadeh, R., Phithakkitnukoon, S., Veloso, M. and Bento, C. 2013. Mining temporal patterns of transport behaviour for predicting future transport usage. *In Proceedings of the 2013 ACM Conference on Pervasive and Ubiquitous Computing*, pp. 1239-1248.

Foell, S., Phithakkitnukoon, S., Kortuem, G., Veloso, M. and Bento, C. 2014. Catch me if you can: Predicting mobility patterns of public transport users. *In 2014 IEEE 17th International Conference on Intelligent Transportation Systems (ITSC)*, 1995-2002.

Foell, S., Phithakkitnukoon, S., Kortuem, G., Veloso, M., & Bento, C. 2015. Predictability of public transport usage: A study of bus rides in Lisbon, Portugal. *IEEE Transactions on Intelligent Transportation Systems*. doi:10.1109/TITS.2015.2425533.

Frumin, M.S. 2010. Automatic data for applied railway management: passenger demand, service quality measurement, and tactical planning on the London Overground Network (*MS Thesis, Massachusetts Institute of Technology*).

Fu, Q., Liu, R. and Hess, S. 2014. A Bayesian modelling framework for individual passenger's probabilistic route choices: A case study on the London underground. *In Transportation Research Board 93rd Annual Meeting* (No. 14-5328).

General Transit Feed Specification (GTFS) 2015. https://developers.google.com/transit/ Accessed 10/11/2015.

Gordillo, F. 2006. The value of automated fare collection data for transit planning: An example of rail transit OD matrix estimation (*MS Thesis, Massachusetts Institute of Technology*).

Gordon, J.B. 2012. Intermodal passenger flows on London's public transport network: automated inference of full passenger journeys using fare-transaction and vehicle-location data (*MS Thesis, Massachusetts Institute of Technology*).

Gordon, J., Koutsopoulos, H., Wilson, N. and Attanucci, J. 2013. Automated inference of linked transit journeys in London using fare-transaction and vehicle location data. *Transportation Research Record: Journal of the Transportation Research Board*, (2343), pp. 17-24.

He, L., Nassir, N., Trépanier, M. and Hickman, M. 2015. Validating and calibrating destination estimation algorithms for public transport smart card fare collection systems. *Report CIRRELT*- pp. 2015-52.

Hofmann, M. and O'Mahony, M. 2005. Transfer journey identification and analyses from electronic fare collection data. *In Proceedings of the IEEE Intelligent Transportation Systems Conference*, pp. 34-39.

Hong, S.P., Min, Y.H., Park, M.J., Kim, K.M. and Oh, S.M. 2015. Precise estimation of connections of metro passengers from smart card data. *Transportation*, pp. 1-21.

Jang, W. 2010. Travel time and transfer analysis using transit smart card data. *Transportation Research Record: Journal of the Transportation Research Board*, (2144), pp. 142-149.

Ji, Y. 2011. Distribution-based approach to take advantage of automatic passenger counter data in estimating period route-level transit passenger origin-destination flows: Methodology development, numerical analyses and empirical investigations (*PhD Thesis, The Ohio State University*).

Ji, Y., Mishalani, R., McCord, M. and Goel, P. 2011. Identifying homogeneous periods in bus route origin-destination passenger flow patterns from automatic passenger counter data. *Transportation Research Record: Journal of the Transportation Research Board*, (2216), pp. 42-50.

Kieu, L.M., Bhaskar, A. and Chung, E. 2014. Passenger segmentation using smart card data. *IEEE Transactions on Intelligent Transport Systems*, doi:10.1109/TITS.2014.2368998.

Kieu, L.M., Bhaskar, A. and Chung, E. 2015. A modified density-based scanning algorithm with noise for spatial travel pattern analysis from smart card AFC data. *Transportation Research Part C: Emerging Technologies*. doi:10.1016/j.trc.2015.03.033.

Kuhlman, W. 2015. The construction of purpose-specific OD matrices using public transport smart card data (*Doctoral dissertation, TU Delft, Delft University of Technology*).

Kurauchi, F., Schmöcker, J.D., Shimamoto, H. and Hassan, S.M. 2014. Variability of commuters' bus line choice: An analysis of oyster card data. *Public Transport*, 6(1-2), pp. 21-34.

Kusakabe, T. and Asakura, Y. 2014. Behavioural data mining of transit smart card data: A data fusion approach. *Transportation Research Part C: Emerging Technologies*, 46, pp. 179-191.

Kusakabe, T., Iryo, T. and Asakura, Y. 2010. Estimation method for railway passengers' train choice behaviour with smart card transaction data. *Transportation*, 37(5), pp. 731-749.

Lathia, N., Froehlich, J. and Capra, L. 2010. Mining public transport usage for personalized intelligent transport systems. *In 2010 IEEE 10th International Conference on Data Mining (ICDM)*, pp. 887-892.

Lathia, N., Smith, C., Froehlich, J. and Capra, L. 2013. Individuals among commuters: Building personalized transport information services from fare collection systems. *Pervasive and Mobile Computing*, 9(5), pp. 643-664.

Lee, M. and Sohn, K. 2015. Inferring the route-use patterns of metro passengers based only on travel-time data within a Bayesian framework using a reversible-jump Markov chain Monte Carlo (MCMC) simulation. *Transportation Research Part B: Methodological*, 81, pp. 1-17.

Lee, S.G. and Hickman, M. 2011. Travel pattern analysis using smart card data of regular users. *In Transportation Research Board 90th Annual Meeting* (No. 11-4258).

Lee, S.G. and Hickman, M. 2013. Are transit trips symmetrical in time and space? Evidence from the Twin Cities. *Transportation Research Record: Journal of the Transportation Research Board*, (2382), pp. 173-180.

Lee, S.G. and Hickman, M. 2014. Trip purpose inference using automated fare collection data. *Public Transport*, 6(1-2), pp. 1-20.

Lee, S.G., Hickman, M. and Tong, D. 2013. Development of a temporal and spatial linkage between transit demand and land-use patterns. *Journal of Transport and Land Use*, 6(2), pp. 33-46.

Li, D., Lin, Y., Zhao, X., Song, H. and Zou, N. 2011. Estimating a transit passenger trip origin-destination matrix using automatic fare collection system. *In Database Systems for Advanced Applications*, pp. 502-513.

Lianfu, Z., Shuzhi, Z., Yonggang, Z. and Ziyin, Z. 2007. Study on the method of constructing bus stops OD matrix based on IC card data. *In, International Conference on Wireless Communications, Networking and Mobile Computing*, WiCom 2007, pp. 3147-3150.

Ma, X. and Wang, Y. 2014. Development of a data-driven platform for transit performance measures using smart card and GPS data. *Journal of Transportation Engineering*, 140(12), 04014063.

Ma, X., Wu, Y. J., Wang, Y., Chen, F. and Liu, J. 2013. Mining smart card data for transit riders' travel patterns. *Transportation Research Part C: Emerging Technologies*, 36, pp. 1-12.

Medina, S. and Erath, A. 2013. Estimating dynamic workplace capacities by means of public transport smart card data and household travel survey in Singapore. *Transportation Research Record: Journal of the Transportation Research Board*, (2344), pp. 20-30.

Morency, C., Trépanier, M. and Agard, B. 2007. Measuring transit use variability with smart-card data. *Transport Policy*, 14(3), pp. 193-203.

Munizaga, M., Devillaine, F., Navarrete, C. and Silva, D. 2014. Validating travel behaviour estimated from smart card data. *Transportation Research Part C: Emerging Technologies*, 44, pp. 70-79.

Munizaga, M.A. and Palma, C. 2012. Estimation of disaggregate multimodal public transport origin–destination matrix from passive smart card data from Santiago, Chile. *Transportation Research Part C: Emerging Technologies*, 24, pp. 9-18.

Nassir, N., Hickman, M. and Ma, Z.L. 2015. Activity detection and transfer identification for public transit fare card data. *Transportation*, 42(4), pp. 683-705.

Nassir, N., Khani, A., Lee, S., Noh, H. and Hickman, M. 2011. Transit stop-level origin-destination estimation through use of transit schedule and automated data collection system. *Transportation Research Record: Journal of the Transportation Research Board*, (2263), pp. 140-150.

Nishiuchi, H., King, J. and Todoroki, T. 2013. Spatial-temporal daily frequent trip pattern of public transport passengers using smart card data. *International Journal of Intelligent Transportation Systems Research*, 11(1), pp. 1-10.

Nunes, A.A., Galvao Dias, T. and Falcao e Cunha, J. 2015. Passenger journey destination estimation from automated fare collection system data using spatial validation. *IEEE Transactions on Intelligent Transportation Systems*, doi:10.1109/TITS.2015.2464335.

Park, J., Kim, D.J. and Lim, Y. 2008. Use of smart card data to define public transit use in Seoul, South Korea. *Transportation Research Record: Journal of the Transportation Research Board*, (2063), pp. 3-9.

Paul, E.C. 2010. Estimating train passenger load from automated data systems: application to London Underground (*MS Thesis, Massachusetts Institute of Technology*).

Pelletier, M.P., Trépanier, M. and Morency, C. 2011. Smart card data use in public transit: A literature review. *Transportation Research Part C: Emerging Technologies*, 19(4), pp. 557-568.

Reddy, A., Lu, A., Kumar, S., Bashmakov, V. and Rudenko, S. 2009. Entry-only automated fare-collection system data used to infer ridership, rider destinations, unlinked trips, and passenger miles. *Transportation Research Record: Journal of the Transportation Research Board*, (2110), pp. 128-136.

Robinson, S., Narayanan, B., Toh, N. and Pereira, F. 2014. Methods for pre-processing smart card data to improve data quality. *Transportation Research Part C: Emerging Technologies*, 49, pp. 43-58.

Schmöcker, J.D., Shimamoto, H. and Kurauchi, F. 2013. Generation and calibration of transit hyperpaths. *Transportation Research Part C: Emerging Technologies*, 36, pp. 406-418.

Seaborn, C., Attanucci, J. and Wilson, N. 2009. Analyzing multimodal public transport journeys in London with smart card fare payment data. *Transportation Research Record: Journal of the Transportation Research Board*, (2121), pp. 55-62.

Spurr, T., Chapleau, R. and Piché, D. 2014. Use of subway smart card transactions for the discovery and partial correction of travel survey bias. *Transportation Research Record: Journal of the Transportation Research Board*, (2405), pp. 57-67.

Sun, L., Lee, D.H., Erath, A. and Huang, X. 2012. Using smart card data to extract passenger's spatio-temporal density and train's trajectory of MRT system. *In Proceedings of the ACM SIGKDD International Workshop on Urban Computing*, pp. 142-148.

Sun, L., Lu, Y., Jin, J.G., Lee, D.H. and Axhausen, K.W. 2015. An integrated Bayesian approach for passenger flow assignment in metro networks. *Transportation Research Part C: Emerging Technologies*, 52, pp. 116-131.

Sun, Y. and Schonfeld, P.M. 2015. Schedule-based rail transit path-choice estimation using automatic fare collection data. *Journal of Transportation Engineering*, doi:10.1061/(ASCE) TE.1943-5436.0000812.

Tamblay, S., Galilea, P., Iglesias, P., Raveau, S. and Muñoz, J.C. 2015. A zonal inference model based on observed smart-card transactions for Santiago de Chile. *Transportation Research Part A: Policy and Practice*, doi:10.1016/j.tra.2015.10.007.

Tao, S., Corcoran, J., Mateo-Babiano, I. and Rohde, D. 2014a. Exploring bus rapid transit passenger travel behaviour using big data. *Applied Geography*, 53, pp. 90-104.

Tao, S., Rohde, D. and Corcoran, J. 2014b. Examining the spatial–temporal dynamics of bus passenger travel behaviour using smart card data and the flow-comap. *Journal of Transport Geography*, 41, pp. 21-36.

Trépanier, M., Tranchant, N. and Chapleau, R. 2007. Individual trip destination estimation in a transit smart card automated fare collection system. *Journal of Intelligent Transportation Systems*, 11(1), pp. 1-14.

Trépanier, M., Morency, C. and Blanchette, C. 2009. Enhancing household travel surveys using smart card data. *In Transportation Research Board 88th Annual Meeting* (No. 09-1229).

Utsunomiya, M., Attanucci, J. and Wilson, N. 2006. Potential uses of transit smart card registration and transaction data to improve transit planning. *Transportation Research Record: Journal of the Transportation Research Board*, (1971), pp. 119-126.

Van der Hurk, E., Kroon, L., Maróti, G. and Vervest, P. 2015. Deduction of passengers' route choices from smart card data. *IEEE Transactions on Intelligent Transportation Systems*, 16(1), pp. 430-440.

Wang, W., Attanucci, J.P. and Wilson, N.H.M. 2011. Bus passenger origin-destination estimation and related analyses using automated data collection systems. *Journal of Public Transportation*, 14(4), p. 7.

Yang, S., Wu, Y.J., Marion, B. and Moses, I.E. 2015. Identification of transit fare box data errors: impacts on transit planning. *Public Transport*, pp. 1-17. doi:10.1007/s12469-015-0107-6.

Yuan, N.J., Wang, Y., Zhang, F., Xie, X. and Sun, G. 2013. Reconstructing individual mobility from smart card transactions: A space alignment approach. *In, 2013 IEEE 13th International Conference on Data Mining* (ICDM), pp. 877-886.

Zhao, J. 2004. The planning and analysis implications of automated data collection systems: rail transit OD matrix inference and path choice modeling examples. (*MS Thesis, Massachusetts Institute of Technology*).

Zhao, J., Rahbee, A. and Wilson, N.H.M. 2007. Estimating a rail passenger trip origin-destination matrix using automatic data collection systems. *Computer-Aided Civil and Infrastructure Engineering*, 22(5), pp. 376-387.

Zhao, J., Tian, C., Zhang, F., Xu, C. and Feng, S. 2014. Understanding temporal and spatial travel patterns of individual passengers by mining smart card data. *In, 2014 IEEE 17th International Conference on Intelligent Transportation Systems* (ITSC), pp. 2991-2997.

Zhou, F. and Xu, R.H. 2012. Model of passenger flow assignment for urban rail transit based on entry and exit time constraints. *Transportation Research Record: Journal of the Transportation Research Board*, (2284), pp. 57-61.

Zhu, Y. 2014. Passenger-to-train assignment model based on automated data. (*MS Thesis, Massachusetts Institute of Technology*).

## AUTHOR BIOGRAPHY

**Mark Hickman** is the ASTRA Chair and Professor of Transport Engineering within the School of Civil Engineering at the University of Queensland. He is also the Director of the Centre for Transport Strategy, also within the School of Civil Engineering. He has worked with smart card data from bus service in Minneapolis/St Paul, Minnesota (USA) and from Brisbane, Queensland (Australia). His primary research interest is in understanding and modelling passenger behaviour in public transit networks, for which smart card data are very well-suited.

Chapter 3

# Destination and Activity Estimation

**A. Ali[1] and S. Lee[2,*]**

**ABSTRACT**

Household travel surveys contain, besides travel information, socio-demographic information which plays an important role in analyzing travel dynamics. Disadvantages of such surveys are though that these are costly to conduct, time-consuming and often do not cover more than 2-3% of the population leading to possible biases and errors in reporting. On the other hand, smart card data provide real-time, accurate and detailed on board transactions records of each user, however, only cover a subset of all trips. Further, smart card data do not usually contain socio-demographic information and hence, one of the most important parameter "trip purpose or activity" is missing. This problem is of growing interest and several algorithms to infer the trip purpose are proposed. In this chapter, the discussion is on trip destination and purpose estimation methods and a 1-day Household Person Travel Diary Survey (HHPTD) is utilized to generate a framework for assigning trip purposes to trips. Activities studied are "home", "work", "school", "academy" and "shopping". Decision trees and rule-based models, namely, R-Tree and C50, are used to estimate the trip purpose. HHPTD is utilized to train the datasets and applied on the smart card dataset to calculate the probabilities of different activities. Activity duration, location and start time are the parameters which were analysed for developing the inference framework. The spatial context of this study is Seoul Metropolitan Area (SMA) with a focus on Seoul City. The smart card data analysed in this study dates back to June 11, 2012 and over 28 million transactions were made during that day.

[1,2] Department of Transportation Engineering, University of Seoul, 523 21st Century Building, 163 Seoulsiripdae-ro, Dongdaemun-gu, 130-743 Seoul, South Korea. Email: sjlee@uos.ac.kr

* Corresponding author

## 1. SMART CARD USE IN TRIP DESTINATION AND ACTIVITY ESTIMATION

Earlier literature published on the use of smart cards in public transportation planning focused on the estimation of origin-destination (OD) matrices in entry-only systems where users only swipe their cards during boarding a transit station or vehicle. Barry et al. (2002) developed a method to estimate station to station OD matrices for New York Metro by assuming that a high percentage of users return to the destination station of their previous trip to start a new trip and by further assuming that a high percentage of users end their last trip of the day where they begin their first trip of the day. Zhao et al. (2007) carried out similar work and proposed a method to infer rail passenger trip OD matrices in the Chicago transit system and examined rail to bus transfers which were ignored in former studies. Farzin (2008) applied global positioning data (GPS traces) to find the location of buses for assigning origin zones and to develop automated bus OD matrices by integrating automatic vehicle location and fare collection system in Sao Paulo, Brazil. Munizaga and Palma (2012), achieved a success rate of 80% for alighting point estimation for a multimodal public transport OD matrix generation in Santiago.

Recent literature on utilizing smart cards is focused on analyzing travel behaviour and service planning of public transport systems. Morency et al. (2007) used the smart card data for measuring performance of transit network over a range of spatial and temporal resolutions. Performance measures studied were vehicle-kilometres, vehicle-hours, commercial speed, passenger-kilometres, passenger travel time and average trip length (see Chapter 9 for a summary and advances on this line of work). They also created run load profiles for transit lines and calculated occupancy level of vehicles. Trepanier et al. (2007) segregated users into different behavioural groups or clusters based on their regularity and daily patterns. Results proposed that data mining techniques help to identify and characterize market segments among the transit users.

Using data from Seoul, Park et al. (2008) studied the reliability of smart card data and its potential to define user characteristics like bus runs by mode, user types, boarding time and travel time distribution for all transit modes. Jang (2010) examined travel time and transfer points for system improvement in Seoul City and thoroughly analysed transfer points over a wide range of trip patterns according to mode.

Sun et al. (2012) recently carried out work in a Singapore context where the use of smart card data took place to extract passenger's spacio-temporal density to illustrate train trajectories. The model also predicts the location of a certain train and the number of on board passengers. A further recent study using smart card data is by Arana et al. (2014) who looked into the impact of weather conditions on transit conditions.

More closely to this study, recently, there has been a growing interest among researchers to use smart card data within an activity-based perspective. Activity-based modelling (ABM) is not a new approach to travel demand analysis, it has been in practice since at least the last decade. This approach answers many behavioural policy measures which cannot be simply answered by traditional four-step modelling techniques. ABM views travel as a demand, a demand from the need to pursue "activities" in space and time. It employs time-use surveys for analysis or space-time prism. Devillaine et al. (2012) developed rules based on travel diary surveys to detect activities of smart card users in Gatineau and Santiago. Purpose assignment criteria were designed to identify weekday activities along with a set of heuristic rules. They studied work, study, home and other activities. Kusakabe and Asakura (2014) developed a data fusion method of smart card data with person trip survey data and estimated the activity purpose with a success rate of 86.2% using the Naïve Bayes probabilistic model (see also Chapter 5 in this book). Lee and Hickman (2014) built a series of heuristic rules for trip purpose assignment by using cluster analysis by observing users' spatial-temporal interactions over the weekdays. Activities were classified into different clusters according to first and last transactions observed during the day. Trips were associated with work, school or other activities. More recently, Yang et al. (2015) proposed a spatial temporal activity preference model by exploiting the data sets from social networking mobile applications based on locations. Development of a fusion framework, to combine the spatial and temporal activity preference model for activity preference inference and applied tensor factorization models. Nassir et al. (2015) proposed off-optimality concepts to improve the accuracy of short activity detection to estimate passengers' true origins and destinations. Short or hidden activities which are often labelled as transfers were studied in detail based on estimation methods which include variables like alternative paths and routes, service headway, walk times and transfer points.

## 2. SMART CARD DATA STRUCTURE IN SEOUL

In 2004, Seoul Metropolitan Government introduced a new distance-based fare collection system called "T-Money", where users swipe their cards for both entering and disembarking the system; however they do not need to confirm the cards during transfers within the subway system. The system is fully integrated and allows up to 4 free transfers if the time span between previous trip segment's alighting and next trip segment's boarding is less than 30 minutes. "Integrated" further refers to users being able to transfer free of charge between different modes such as bus and train. Each time a transaction is made the smart card system records individual transaction information for entry and exit and creates trip-based records.

The recorded information on smart cards (Table 1) reveals detailed information on a user's complete day itinerary. Unique card ID, boarding

**Table 1.** Information stored in smart card database per transaction

| Information | Description |
|---|---|
| Card ID | Card number for each smart card |
| Departure time | Departure time |
| Type of mode | Bus (local/main/feeder/metropolitan/circle bus), Metro |
| Number of transfers | Number of transfers (from 0 to 4) |
| Type of user | Youth (>12 & < 20y), Children (<12 y) or Adult (>20 y) |
| Boarding time | Boarding time (year/month/day/hour/minute/second) |
| ID of boarding location | Given number of boarding bus/metro stop |
| Alighting time | Alighting time (year/month/day/hour/minute/second) |
| ID of alighting location | Given number of alighting bus/metro stop |
| Number of passengers | Number of passengers |
| Basic fare | Starting (base) fare |
| Additional fare | Additional fare with distance |
| Travel distance | Distance from origin stop to destination stop |

time, boarding station ID, alight time, alight station ID, transport method (subway, regional bus, circular bus, etc.), bus route ID, passenger type (adult, youth, children), total fare and total distance travelled are the attributes recorded in the smart card database for each singular trip segment pair. Therefore, it is necessary to distinguish between trip legs and trips as explained later in the method section. To date, over 100 million prepaid cards have been issued with 71 million affiliated cards in use and about 30 million transactions were made in greater SMA every day. The smart card penetration rate in SMA is 92% (2013) and is on constant rise (Figure 1). Therefore, the data can accurately generate the transit-demand due to availability over a large period.

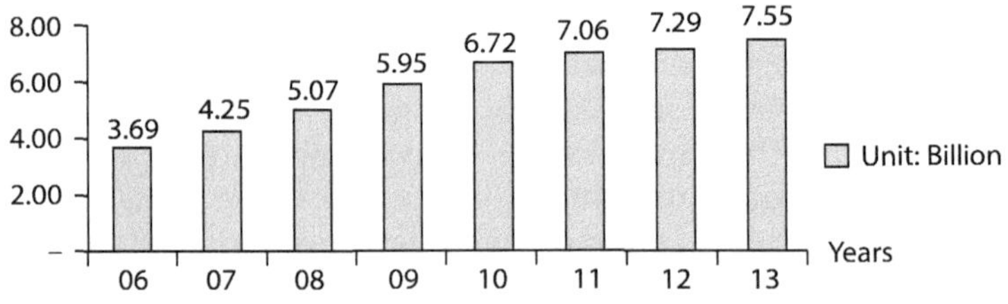


**Fig. 1.** Number of trips using smart card payment (Korea Smart Card Corporation)

The public transport system comprises 19 urban railway lines and more than 400 bus routes in greater SMA. Public transport accounts for about 63% of daily trips in greater metropolitan area (Seoul Statistics, 2013[1]).

[1] Composition of Daily Passenger Transportation. Could be accessed at: http://english.seoul.go.kr/

# 3. METHODOLOGY FOR TRIP DESTINATION ESTIMATION

## 3.1 Data Cleaning

The raw database consists of 18.83 million trip segment records (storage of each transaction as an independent OD trip). The total number of daily passengers is 6.6 million, each with 2.82 average transactions per day. The daily average number of trips per person is 2.11 and about 1.2 million users make only one trip per day (one transaction per card). For analysing activities, at least 2 trips are in need (or transactions). These users are regarded as non-frequent metro users and need to be discarded for activity analysis because they do not use the metro for the complete journey. For analysing activities, at least 2 transactions are needed to get information on duration of the activity. Therefore, the users with one transaction are trimmed off from the database. Users with more than 9 transactions can also be trimmed off since the less than 0.5% of the users and using them the in analysis will complicate the activity purpose imputation process. Listed are following assumptions (similar to Lee and Hickman, 2014) for data trimming:

- The origin of the first trip is also the destination of the last trip of the day (activity type is "home").
- The trip destination is also the origin of the succeeding trip (search radius of 700 metres is fixed in this case study because of densely located transit stations in Seoul).
- Transit users do not switch to other transport modes within their given sequence of daily transit trips.

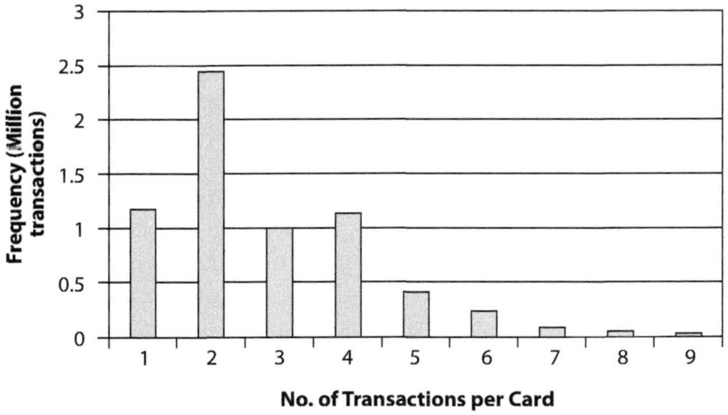


**Fig. 2a.** Summary statistics of smart card data

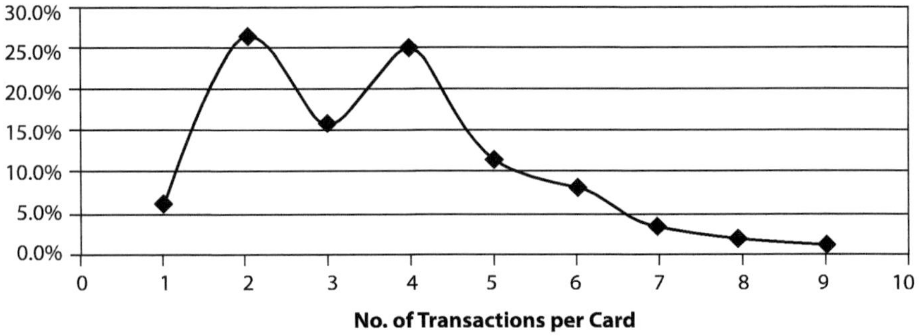


**Fig. 2b.** Percentage over total transactions

Another implication is the distance between consecutive boarding and alighting points (2nd assumption). If the trip boarding point of the current trip segment (or a trip) is far away from the alighting point of previous trip segment (or trip), then it indicates that the user has switched to an unidentified mode of transportation (bike, car share, walk, etc.) for his next trip. A threshold of maximum walking distance to be less than 1 km is set in this study. It is unusual that a user would walk more than 1 kilometre for his next trip boarding in such a connected transit (bus and subway) network with highly densely located stops. The data analysis indicates that around 62% of the users start their next trip from the same (subway) station where they had alighted in previous trip. Another 30% of the users start their next journey in proximity of 700 metres to the previous trip's alighting location. The remaining 8% of the users start their next trip, on average, almost 6.5 km away from the previous trip alighting point indicating an unidentified mode shift (taxi, bicycle or car-pooling).

## 3.2 Trips and Trip Legs

A trip might be composed of several trip legs (here referred to as individual transaction) and/or different modes; therefore, it is mandatory to distinguish between trips and its legs (Figure 3). Data kept as an individual trip segment in the smart card database and the transfer points are distinguished according to the maximum allowable transfer time which is 30 minutes for Seoul (Ali et al. 2015). A program is written in java language which reads individual smart card transactions and gives the output as a trip, after which an activity with, at this stage of the analysis, unknown purpose is performed. Other attributes included are activity start and end time, duration and location. The last transaction (alighting) of the day is then considered the home location if the first trip's boarding and last trip's alighting locations are not farther than 500 metres.

For each trip, if the time difference between current trip's alighting ($D_i$) and next trip's boarding ($O_{i+1}$) is greater than 30 minutes, it is inferred as an activity, otherwise a transfer point. In the next section, R Tree and C50

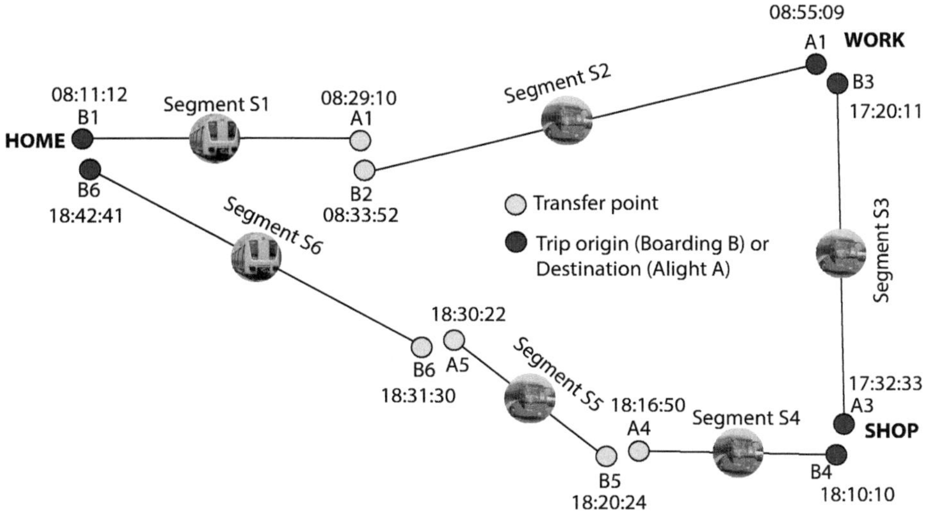


**Fig. 3.** Complete day itinerary of a user (Source: Ali et al., 2015)

algorithms are used to predict the trip purpose which is unknown at this stage.

## 4. TRIP PURPOSE IMPUTATION USING HOUSEHOLD TRAVEL SURVEY

The Korean Transport Database (KTDB) conducted the 2010 HHPTD survey and the survey includes one-day trip information about each household. The data is composed of 217,444 households with 540,298 persons surveyed. The average trip rate per person is 2.46 for private car users compared to 2.07 for public transport users, which is very similar to the smart card data (i.e., 2.11). For each household member, the variables recorded are: person unique ID, trip purpose, trip mode, departure time, arrival time, activity duration and sequence number. Activities recorded in the original survey includes "home", "work", "school", "academy", "work-based trip", "shopping", "leisure" and "others" (Table 2). Work trips are defined as the trips made during the day for work purpose which could be both home-based and work-based. "Home-based trips" refer to the trips which are organized with home as either the origin or the destination of the trip. Similarly, work-based trips refer to the business trips made from workplaces which include business related travel. School trips are defined as the trips made during the day for educational purposes (schools) where academy trips (which are also reported in the Household Travel Survey) are the trips made during afternoon and evening periods for private tuitions. Shopping includes buying groceries in the market. 18 modes of transport are recorded in which "walk", "private car", "car pool", "bus", "subway", "railway", "taxi", "motorcycle" and "bicycles" are the dominant ones. Based on trip purpose, 96% of the trips are home-based while 4% of the trips are work-based trips. Out of those 96% home-based trips, 45.7% activities are "home", 19.5% "work", 11.9% "school", 5.3% "academy", 2.3% "shopping", 3.7% "leisure" and 6.1% "others". Transit users are extracted from the data and trip characteristics of public transport users to develop a framework for trip purpose inference process. Figure 4 shows the cumulative density functions (CDF) of activity start time and activity duration for each activity, based on HHPTD. The CDFs will be used to identify activity types in the smart card database.

**Table 2.** Reported activities in household travel survey

| Activity Type | Percentage |
|---|---|
| Home | 45.7% |
| Work | 19.5% |
| School | 11.9% |
| Academy | 5.3% |
| Shopping | 2.3% |
| Leisure | 3.7% |
| Others | 6.1% |
| Work-based travel | 4% |

## 4.1 Activity Start Time and Duration

Figure 4 shows the activity start time regimes for home, work, study (school and academy), shopping and others. The data interval is 30 minutes which helps to illustrate that majority of the work activities start between 6:30 am and 9:30 am (work start time peak – red coluor). Approximately 8% of the total activities during 8:00 – 8:30 am and 8:30 – 9:00 am are work activities. The work activity graph sharply peaks at 6:30 am in the morning and abruptly declines after 09:30 am. The work activities graph again rises during afternoon period indicating work-based trips and lunch break trips (less than 0.5% people returned home/travelled for lunch breaks in the survey).

For home returning trips, it is quite visible that home activities start gradually from the afternoon period indicating students returning to homes from schools and colleges. The graph peaks up in the evening strip (PM peak) corresponding to high school students returning from colleges/universities and workers returning from job places (6:00 pm is the earliest finish time in most Korean companies). The slope gradually decreases while still 12% of the total home-return trips are seen between 9:00 – 10:00 pm which corresponds to the usual late night working habit in South Korea. Similarly, study trips start in the morning peak, i.e., schools, colleges and universities where academy trips start mostly in the evening period.

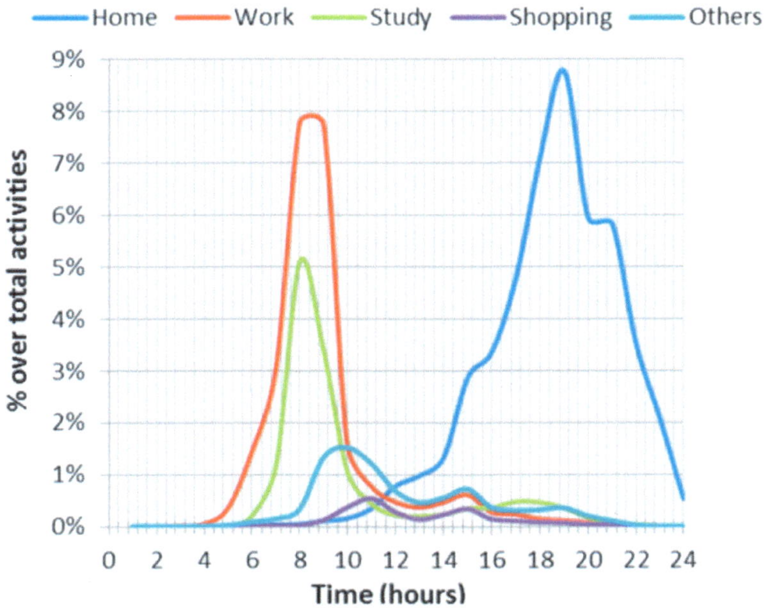


**Fig. 4.** Activity start time regimes (HHPTD, Public transport users only)

Activity start time and duration are the parameters used to train the data sets in R packages. The training data sets are then used to predict the activity purposes. Based on the CDF, most of the work activities have about 650 minutes of duration with a standard deviation of 110 minutes. Similarly, the average duration for school, academy and shopping activities comes out to 5, 2 and 1 hour.

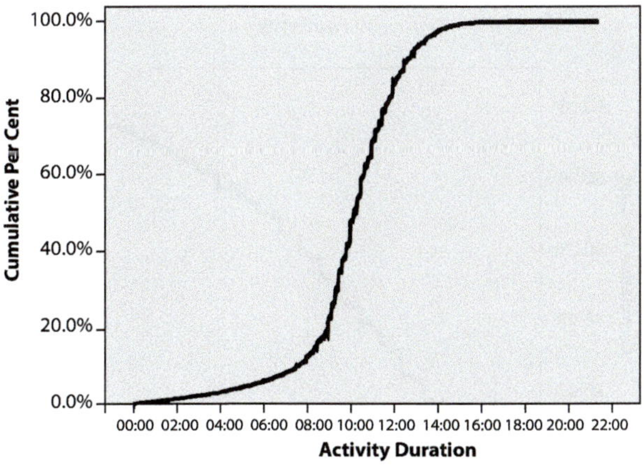


(a) Work

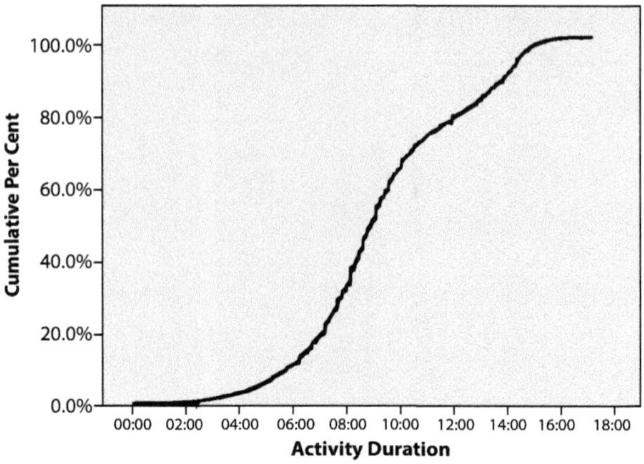


(b) School

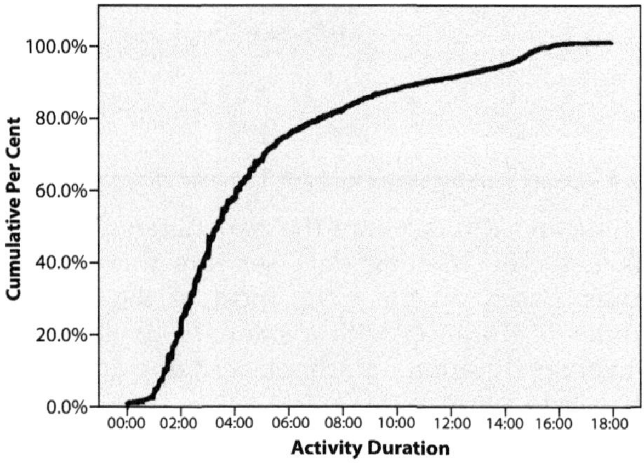


(c) Academy

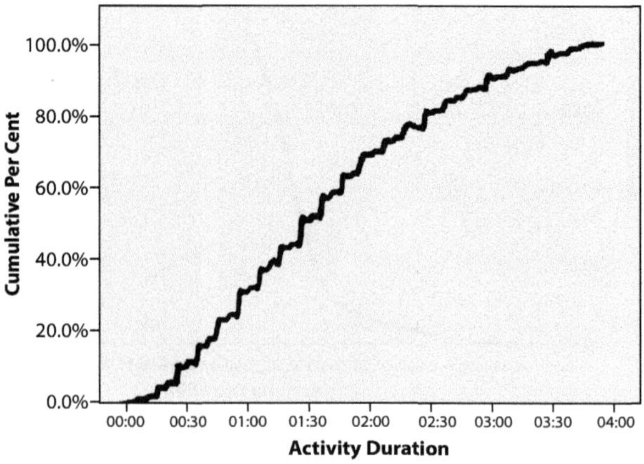


(d) Shopping

**Fig. 5.** CDF of activity duration (HHPTD)

## 4.2 Trip Purpose Prediction

Restructuring of household person travel diary survey data and each activity represented by these attributes: person ID (household), activity start time, activity duration and activity location. Here, the activity purpose is already known and models are applied Tree and C50 models in this section to predict the unknown activities from the smart card data and calculate the misclassification error. 50% of the household data is used to train the model and the remaining 50% is used to calculate the misclassification error. Home activities from household results are not used for prediction purposes since it is assumed, as discussed in Section 3.1, that the origin of the first trip is destination of the last trip of the day. Activity types such as work, academy, use of shopping and others are used to calculate the probabilities of trip purpose based on activity start time, activity duration and location.

### *4.2.1 Tree Classification*

Tree classification algorithms[2] written in R language are used to predict the activity purpose here. The basic decision tree models (or decision support tools) are either classification trees, applicable to binary response variables, or regression tree models, applicable to numeric response variables. The tree model assigns every record in a data set to a unique group and generated a predicted response for each group. The smart card data is designed as depicted in Table 3a and b, while data from the household travel diary survey also contain a similar structure plus the additional information of activity type.

At the start, a distinction is made between 2 smart card types, i.e., adults (>20 years) and youth (<20 years). Dominating activity for adults is "work" and for children it is "school" and "academy".

**Table 3a.** Trip chain of a typical smart card user (1-person)

| Smart Card ID | Activity Start Time | Activity Duration (Hours) | Location (Sub District) |
|---|---|---|---|
| 1120092190 | 08:22:13 | 9 | 1101054 |
| 1120092190 | 17:45:12 | 2 | 1101057 |
| 1120092190 | 21:20:31 | Last transaction (alight) of the day: regarded as home | 1101054 |

**Table 3b.** Activity chain from household travel survey (1-person)

| Person ID | Activity Start | Activity Duration (Hours) | Location (Sub District) | Activity Purpose |
|---|---|---|---|---|
| 1212450001 | 08:40 | 9 | 1101071 | Work |
| 1212450001 | 18:30 | 1 | 1101057 | Shopping |
| 1212450001 | 20:00 | Last activity | 1101071 | Home |

[2] https://cran.r-project.org/web/packages/tree/tree.pdf

In the tree classification model, two predictors, namely, activity start time and duration are used. The overall predicted misclassification error comes out to 0.28 for the tree classification model which means that 28% of the activities were not estimated correctly. The main source of error is that there are many overlapping activities if the estimation is solely based on two parameters only (activity start and duration). Work, school and academic activities are estimated with a success rate of 81% because the start time and duration for these activity types follow regular patterns, whereas activity types such as shopping and others do not follow regular patterns and hence have many communalities accounting for misclassification. The test results are as shown in Table 4.

### 4.2.2 C50 Algorithm

Decision tree as well as the rule-based model C50[3] are further used to calculate the probabilities of activities based on the given three factors: activity start time, duration and location. The C50 algorithm is widely used as a decision tree method in machine learning. The prediction power and efficiency of C50 (which is a modified form of C4.5) is greater than simple tree algorithm. C50 builds the decision trees based on concepts of information entropy. The training data is a set to already classified samples. Overall misclassification based on three parameters using C50 comes out to 0.20, meaning 80% success rate for prediction. The third parameter is activity location which improves the overall prediction results, since choice of activity location depends upon the land-use information.

## 5. RESULTS AND DISCUSSION

The results in Tables 4a and b shows the cross validation summary for the two algorithms used in this study. The overall misclassification error using the two variables as predictor (activity start time and duration) in R-Tree classification algorithm comes out to 0.28. Work and school activities have a stringent start times and most follow typical patterns in terms of start time and duration (9 hours – average work duration). HHPTD reports 4% of total trips to be work-based trips and these are not considered separate activities to minimize misclassification error. Further, the household survey data report that the majority of work-based trips are carried out not with public transport but by car or walk. As the prediction is solely based on duration and start time, there are many overlaps with activity types "shop" and "other".

Similarly, the school activity is also misclassified up to 16% (among activity types "academy", "shop" and "other"). The majority of the trips with purpose "academy" starts in the afternoon and evening time regime

---

[3] https://cran.r-project.org/web/packages/C50/C50.pdf

**Table 4a.** Cross validation summary for training data using R-Tree

| | Predicted Group | | | | | |
|---|---|---|---|---|---|---|
| | **Work** | **School** | **Academy** | **Shop** | **Others** | **Total** |
| **Work** | 0.81 | 0 | 0 | 0.09 | 0.1 | 1 |
| | 81% | 0% | 0% | 9% | 10% | 100% |
| **School** | 0 | 0.84 | 0.07 | 0.03 | 0.06 | 1 |
| | 0% | 84% | 7% | 3% | 6% | 100% |
| **Academy** | 0 | 0.02 | 0.71 | 0.12 | 0.15 | 1 |
| | 0% | 2% | 71% | 12% | 15% | 100% |
| **Shop** | 0.11 | 0.09 | 0.09 | 0.61 | 0.1 | 1 |
| | 11% | 9% | 9% | 61% | 10% | 100% |
| **Others** | 0.08 | 0.05 | 0.13 | 0.15 | 0.59 | 1 |
| | 8% | 5% | 13% | 15% | 59% | 100% |

**Table 4b.** Cross validation summary for training data using C50 algorithm

| | Predicted Group | | | | | |
|---|---|---|---|---|---|---|
| | **Work** | **School** | **Academy** | **Shop** | **Others** | **Total** |
| **Work** | 0.89 | 0 | 0 | 0.05 | 0.06 | 1 |
| | **89%** | 0% | 0% | 5% | 6% | 100% |
| **School** | 0 | 0.88 | 0.03 | 0.02 | 0.07 | 1 |
| | 0% | **88%** | 3% | 2% | 7% | 100% |
| **Academy** | 0 | 0.02 | 0.81 | 0.07 | 0.1 | 1 |
| | 0% | 2% | **81%** | 7% | 10% | 100% |
| **Shop** | 0.06 | 0.07 | 0.05 | 0.74 | 0.08 | 1 |
| | 6% | 7% | 5% | **74%** | **8%** | 100% |
| **Others** | 0.05 | 0.03 | 0.11 | 0.12 | 0.69 | 1 |
| | 5% | 3% | 11% | 12% | **69%** | 100% |

corresponding to a total prediction rate of 71% with 29% misclassified into activity types "shop" and "other" due to overlaps in duration and start time. Shopping activities do not follow any general trend in terms of start time and duration. The overall prediction rate increases by taking three variables into account with activity location being the third predictor. There would be a high possibility for the trip purpose being "shop" if the bus stop or metro station is at or near a shopping mall. Similarly, there would be a high possibility of the trip purpose to be leisure, if the location is at or near the sports complex or a park. In this study, the activities like "leisure" and work-based trips were not considered as this would complicate the overall prediction process and increase the misclassification error. Deriving activity chains and validating it against the HHPTD would be the next step because there is a unique card ID for each person and the activity chains could be easily derived then.

## 6. ILLUSTRATION OF RESULTS WITH MATSim

Detection of activities of transit smart card users has many implications on analyzing the travel patterns. Predicted work activities are converted into the input demand file of a simulation package, MATSim[4] (Multi-Agent Transport Simulations). Commuting trips are then simulated and work locations of typical bus line users are visualized. Figure 6a and 6b show the home and work locations of users of bus route number 420. As it is evident from Figure 6b, most of the users work in the areas where they have to transfer to other transit lines. Those working in the central areas of Seoul above Han River would transfer to other routes to reach their work places. Such information can help practitioners and transit authorities to obtain an insight on the location of the potential users and improve their services by adding more direct routes that minimize transfers and increase the level of service.

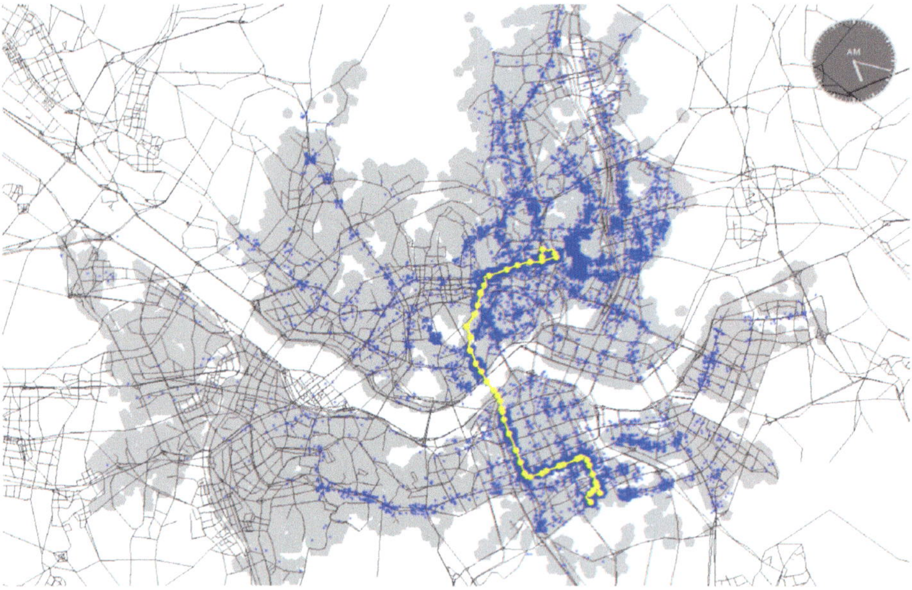

(a) Home location

[4] http://www.matsim.org/

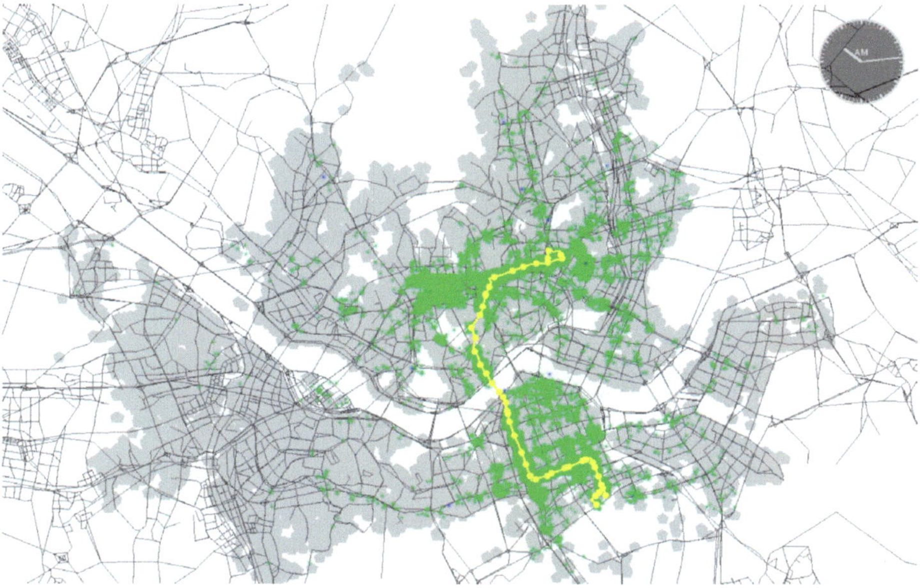


(b) Work location

**Fig. 6.** Home and work locations of a typical bus line user (route 420)

## 7. CONCLUSION

The focus in this chapter was on methods for estimating destination and activities related to smart card users. Our experience is that the main activity types such as home, work and educational activities which include both school and academy could be predicted using various models. Previous researchers have applied various models such as naïve Bayes probabilistic models and achieved satisfactory results. In this chapter applications of tree classification and decision tree models are discussed. It is found that the decision tree algorithm C50 performs better. Results of these models could be implemented in various activity-based planning tools and can help operators to understand behaviour patterns and flexibility in demand, which is sensitive to land-use changes. For improvements of prediction results, it would be desirable if household travel surveys could include questions such as number of transfers, route choice as well as more accurate information on travel times and frequency of activities over a weekly horizon. Such improvements would lead to more accurate predictions in various planning tools. The discussion on fusion of household survey and smart card data is continued in Chapter 5.

## REFERENCES

Agard, B., Morency, C. and Trépanier, M. 2007. Mining Public Transport user Behaviour from Smart Card Data. Publication CIRRELT-2007-42. Centre interuniversitaire de recherche sur les reseaux d'entreprise, la logistique et le transport (CIRRELT), Canada. https://www.cirrelt.ca/?Page=DocumentsRecherche2007. Accessed July, 2014.

Ali, A., Kim, J.Y. and Lee, S.J. 2015. Travel behaviour analysis using smart card data. KSCE Journal of Civil Engineering, http://dx.doi.org/10.1007/s12205-015-1694-0.

Arana, P., Cabezudo, M. and Peñalba, M. 2014. Influence of weather conditions on transit ridership: A statistical study using data from smart cards. *Transportation Research Part A*, Vol. 59, pp. 1-12.

Bagchi, M. and White, P.R. 2005. The potential of public transport smart card data. *Transport Policy*, Vol. 12, pp. 464-474.

Barry, J.J., Newhouser, R., Rahbee, A. and Sayeda, S. 2002. Origin and destination estimation in New York City with automated fare system data. *Transportation research record: Journal of the Transportation Research Board*, No. 1817, Transportation Research Board of the National Academies, Washington, D.C., pp. 183-187.

Devillaine, F., Munizaga, M. and Trépanier, M. 2012. Detection of activities of public transport users by analyzing smart card data. *Transportation Research Record*, Vol. 2276, pp. 48-55. Published by Transportation Research Board, Washington.

Farzin, J. M. 2008. Constructing an automated bus origin-destination matrix using farecard and global positioning system data in Sao Paulo, Brazil. *Transportation Research Record: Journal of the Transportation Research Board*, No. 2072, Transportation Research Board of the National Academies, Washington, D.C., pp. 30-37.

Jang, W. 2010. Travel time and transfer analysis using transit smart card data. *Transportation Research Record: Journal of the Transportation Research Board*, No. 2144, Transportation Research Board of the National Academies, Washington, D.C, pp. 142-149.

Kusakabe, T. and Asakura, Y. 2014. Behavioural data mining of transit smart card data: A data fusion approach, *Transportation Research Part C: Emerging Technologies*, Vol. 46, September 2014, pp. 179-191, ISSN 0968-090X, http://dx.doi.org/10.1016/j.trc.2014.05.012.

Lee, Sang Gu. and Hickman, Mark. 2014. Trip purpose inference using automated fare collection data. *Public Transport*, 6 1-2: 1-20. doi:10.1007/s12469-013-0077-5.

Morency, C., Trépanier M. and Agard, B. 2007. Measuring transit performance using smart card data. *Presented at World Conference on Transport Research*, San Francisco, USA.

Munizaga, M.A. and Palma, C. 2012. Estimation of a disaggregate multimodal public transport origin–destination matrix from passive smart card data from Santiago, Chile. *Transportation Research Part C*, Vol. 24, pp. 9-18.

Nassir, N., Hickman, M. and Ma, Z.L. 2015. Activity detection and transfer identification for public transit fare card data. *Transportation*, 42 4: pp. 683-705. doi:10.1007/s11116-015-9601-6.

Park, J.Y., Kim, D.J. and Lim, Y.T. 2008. Use of smart card to define public transit use in Seoul, South Korea. *In Transportation Research Record: Journal of the Transportation Research Board*, No. 2063, Transportation Research Board of the National Academies, Washington, D.C., pp. 3-9.

Sun, L., Lee, D.H., Erath, A. and Huang, X. 2012. Using smart card data to extract passenger's spatio-temporal density and train's trajectory of MRT System. *Published in Proceedings of the ACM SIGKDD International Workshop on Urban Computing*, pp. 142-148.

Trépanier, M., Habib, K.H.N. and Morency, C. 2012. Are transit users loyal? Revelations from a hazard model based on smart card data. *Canadian Journal of Civil Engineering*, Vol. 39, No. 6, pp. 610-618.

Yang, D., Zhang, D., Zheng, V.W. and Yu, Z. 2015. "Modelling user activity preference by leveraging user spatial temporal characteristics in LBSNs. *Systems, Man, and Cybernetics: Systems, IEEE Transactions on*, Vol. 45, no.1, pp. 129, 142, Jan. 2015 doi: 10.1109/TSMC.2014.2327053.

Zhao, J., Rahbee, A. and Wilson, N.H.M. 2007. Estimating a rail passenger trip origin-destination matrix using automatic data collection systems. *Computer-Aided Civil and Infrastructure Engineering*, Vol. 22, pp. 376-387.

## AUTHOR BIOGRAPHY

**Atizaz Ali** has graduated under tutelage of Dr. Seungjae Lee in February 2015. His master's thesis was related to the development of activity-based models in the Seoul Metropolitan Area using smart card data. He also published a research article related to travel behaviour analysis with smart cards in Journal of Korean Society of Civil Engineering. Currently, he holds a position as a researcher at the Singapore-ETH center for Global Environmental Sustainability.

**Dr. Seungjae Lee** has been involved in teaching, research and consultancy in the area of Transportation Planning for the past twenty years in the University of Seoul. He has served, among others, for the Journal of Advanced Transportation as an Editor. He has used smart card data for Seoul Metropolitan Areas to develop activity based modelling.

Chapter 4

# Modelling Travel Choices on Public Transport Systems with Smart Card Data

**S. Raveau**[1,*]

**ABSTRACT**

Understanding and modelling traveller's decisions on public transport systems by correctly analysing their choices and being able to forecast flows on the network are essential elements in urban planning. For this, smart cards have arisen as a valuable information source in the past decade, providing massive information at low-cost. This chapter analyses how smart card data can help us understand traveller's decisions within public transport systems, identifying the relevant factors being taken into account and quantifying the impact that different characteristics of the system have on the preferences of travellers. A case study for Santiago, Chile is presented; the study incorporates perceptions and preferences on a variety of factors (such as crowding, transferring and network topology) to enhance the explanatory and forecasting capabilities of travel demand models.

## 1. INTRODUCTION

Public transport systems play a fundamental role in developing any city. To effectively promote the use of public transport (in contrast to private modes), it is fundamental to understand the decision-making process of public transport travellers, as well as their preferences and perceptions. Within a public transport system, travel decisions are made at two levels: (i) choice of public transport mode (bus, metro, tram, multi-modal trips, etc.) and (ii) choice of travel route (selection of public transport lines and transfer points along the way). There is a dependency in these decisions

[1] Singapore-MIT, Alliance for Research and Technology (Smart). 1 Create Way, #09-02 Create Tower, Singapore, 138602. Email: sraveau@mit.edu

* Corresponding author

and their order might depend on a particular public transport system's characteristics.

Understanding how public transport users make their travel decisions and being able to predict their behaviour is essential in transportation planning. Route choice models have been developed for private transport networks (Bovy and Stern 1990; Ramming 2001; Prato 2009), but not much work has been done in public transport networks (Hunt 1990; Bovy and Hoogendoorn-Lanser, 2005; Raveau et al. 2011). The route choice variables normally included in traditional route choice models limit to some basic service levels attributes of the alternative routes, such as travel time and fare (Ortúzar and Willumsen 2011). However, other variables, related to both the level of service and the traveller's perceptions, influence the user's route choice process but are generally ignore in traditional modelling.

This chapter proposes one main research question: what are the relevant factors that affect multi-modal route choices within a public transport system? The answer to that question determines, at the end, the levels and structure of public transport demand. Any policy maker or practitioner looking to answer that question must face two main tasks: data collection and mathematical modelling. It is on the data collection task where smart cards (and other intelligent transport system technologies) have arisen as a valuable information source in the past decade. Traditional data collection methods used to rely on paper-based surveys, which could be tailor-made for the particular aim of a study, but usually offer small penetration rates at high costs. Smart card data, however, can provide massive information at low-cost (in fact, the most time/resource consuming aspect tends to be related to the processing of the information), but is more rigid in terms of information collected.

The contents of this chapter focus almost exclusively on analysing the travel decisions of public transport users, without analysing the decisions of travellers of other modes (most significantly, car users). Although, the decision of choosing public transport modes over the other alternatives is a significant one, the focus is on analysing the subsequent decisions once the travellers have decided to use the public transport system.

The organization of the reminder of the chapter is as follows. In Section 2, there is a presentation of the theoretical background for analysing and modelling travel decisions on public transport systems; in Section 3, there is discussion on some particularities of modelling multi-modal route choices on public transport systems with smart card data, in Section 4, there is a presentation on a case study for the public transport system of Santiago, Chile; and finally in Section 5, the main conclusions are given.

## 2. THEORETICAL BACKGROUND

This section explores the methodological framework for modelling multi-modal route choice behaviour in public transport systems. The mathematical modelling tools were mostly developed to analyse these decisions on private transport networks (and therefore could be extended to the study of multi-modal networks). The core assumption is that the public transport alternatives could be defined as multi-modal routes (i.e., routes that can incorporate legs in different modes, such as bus and metro) and therefore the focus is on route choice modelling. As stated by Prato (2009), route choice analysis comprises two major modelling challenges: (i) generating a choice set of alternative routes and (ii) estimating a discrete choice model.

A conceptual behavioural framework, which relates both modelling challenges from a behavioural point of view, is shown in Figure 1 (Bovy 2009). A public transport system comprises of a set of *Existing Routes* between different origins and destinations. Considering routes with loops, this set could be potentially infinite. Depending on the levels of available *Information,* a particular traveller can only consider his/her *Known Routes.* Any unknown route will never be selected, independent of its level of service. Depending on individual or external *Restrictions* (such as possessing a public transport pass for a particular mode), the traveller can only choose between his/her *Available Routes.* Obtaining this set of alternative routes corresponds to the first modelling challenge. Depending on his/her *Preferences* (i.e., the relative importance given to the different attributes that include the level of service), the traveller can then *Order* the available routes from most attractive to less attractive. Finally, when the *Decision* is made, the modeller can see the *Chosen Route.* Understating the decision-making process that lead to this choice corresponds to the second modelling challenge.

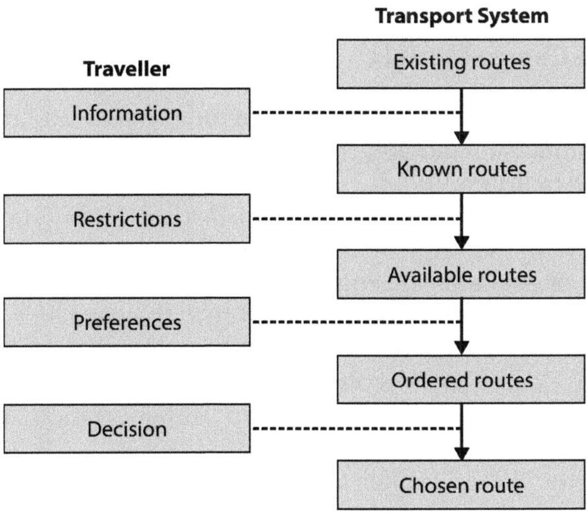


**Fig. 1.** Route choice behaviour
(Adapted from Bovy 2009)

## 2.1 Choice-Set Generation Methods

When understanding and predicting route choice decisions within a transport network, it is necessary to explicitly generate a set of alternative routes (from which the travellers will choose their best alternative). Unlike other discrete choice contexts (particularly, mode choice), transport networks tend to offer a large number of routes between a given origin and destination. Enumerating all these alternatives cannot be done in practice, and at the same time it is well accepted that travellers only consider a subset of all possible alternatives (either because of lack of information, restrictions or preferences). This way, different generation methods have been proposed to generate reasonable choice sets. These methods tend to rely on shortest-paths algorithms, with different specifications of the network costs, and simulation methods. A comprehensive review of some of the existing choice-set generation methods for route choice can be found in Prato (2009).

When defining what constitutes an alternative route, it is necessary to consider the different travel strategies that travellers can follow. In public transport networks with uncertainty on the waiting times (i.e., frequency based systems without timetables), travellers can reduce their expected total travel time by considering a set of common lines. The travellers may not choose a single service, but may board the first service from a set of lines (Chriqui and Robillard 1975; Spiess and Florian 1989). Therefore, an alternative route could be composed of a sequence of multi-service legs. This is a significant difference between route choice modelling in private transport networks and route choice modelling in public transport networks.

## 2.2 Discrete Choice Models

There are different choice models based on system attributes perceived by travellers and their socio-economic and demographic characteristics (in the context of route choice, see for example, Dial 1971; Daganzo and Sheffi 1977; Ramming 2001; Prashker and Bekhor 2004). The basis of these models relies on an assumption of rationality; each traveller chooses a route (among the set of available alternatives) to get the maximum possible utility level (McFadden 1974). It is also assumed that the modeller, who is just an observer, does not have perfect information about the decision-making process, which leads to probabilistic choice models. The most widely used of these discrete choice models (in transportation analysis) is the Multinomial Logit Model (MNL).

Although the MNL model is widely used due to its simplicity, its principal limitation is that it does not consider correlation between alternatives. This might be particularly serious when modelling route choices, as strong correlation between the alternative routes may arise due to overlapping. In urban public transportation networks, the routes linking a given origin-destination pair will typically have many overlaps due to common arcs, so the independent error assumption of the MNL model is unrealistic.

To further understand the problem of route overlapping, let us consider the simple network depicted in Figure 2(a), where there are three alternative routes with the same length $d$, two of which share a common link with length $(d–\alpha)$. For simplicity, we can assume that the length is the only relevant attribute of the utility. If a MNL is applied, the probability of choosing any route (in particular the upper link, which has no overlapping) is 1/3, independent of the value of $\alpha$. This is consistent with the case depicted in Figure 2(b), where $\alpha = d$ and there is no overlapping (the three alternatives are independent, and have a probability of being chosen of 1/3). Nevertheless, if $\alpha = 0$ as shown in Figure 2(c), the two overlapping routes collapse into a single alternative, and the probability of choosing any route (in particular the upper link, which has no overlapping) is 1/2. This way, it is clear that the probability of choosing the upper link varies between 1/3 and 1/2 depending on the value of $\alpha$, something that the MNL model cannot accommodate.

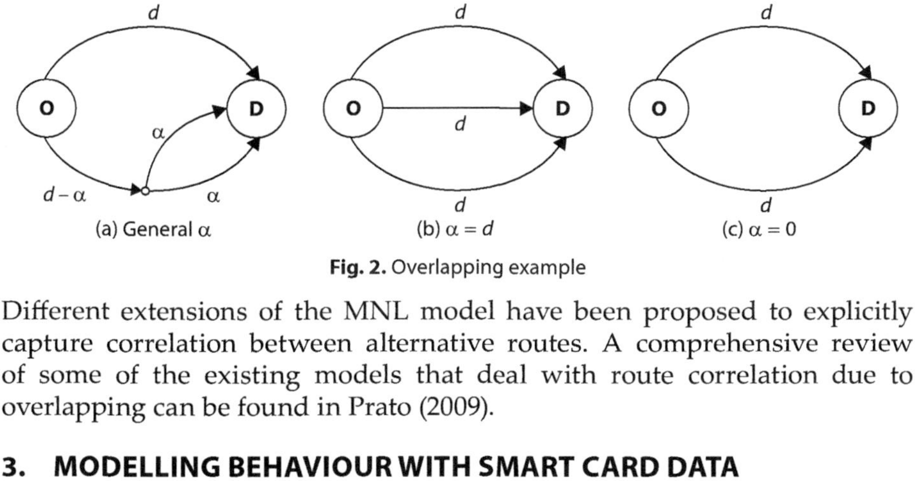


**Fig. 2.** Overlapping example

Different extensions of the MNL model have been proposed to explicitly capture correlation between alternative routes. A comprehensive review of some of the existing models that deal with route correlation due to overlapping can be found in Prato (2009).

## 3. MODELLING BEHAVIOUR WITH SMART CARD DATA

In the literature, it is possible to find multiple route choice models (mainly MNL models or extensions of it), where the utility level of the alternative routes depends on some key route attributes. Usually, these route attributes are all tangible and quite limited to travel time components, fare and transfers. Depending on the available information, the representative utility is sometimes refined using individual socio-economic characteristics, like income or gender, to model different preferences across the travellers. Nevertheless, it is a well-established fact that public transport traveller's decisions are affected by psychological considerations such as aesthetics, comfort and travel-time reliability (see Papinski et al. 2009). However, there are inherent difficulties in integrating this kind of factors into route choice modelling that stem from (i) there subjectivity, given that each user perceives them differently; and (ii) there tangibility, since there is no scale for measuring them.

## 3.1 Modelling Origins and Destinations

A key element when modelling multi-modal route choices in public transport systems is knowing the origin and destination of each trip. That is the input for the first modelling challenge, generating a choice set of alternative routes. In this sense, smart card data provide significant information of origins for each trip leg, especially on systems where its penetration rate is high (as on many public transport systems around the world different payment technologies coexist). Unfortunately, data about destinations is not always available, as some public transport systems with flat fare schemes require the traveller to tap-in (recording origins) but not to tap-out (not recording destinations).

This way, smart card data by itself, in many cases, is not enough to model route decisions in public transport systems. In those cases, additional modelling and/or data processing techniques are necessary. These techniques might require combining different data sources (e.g., smart card data + GPS data), applying mathematical models to infer the destinations, (see Chapter 2 or Munizaga and Palma, 2012), or changing the modelling approach to model sequences of stops (based on the recorded tap-ins) instead of whole routes.

An additional limitation of smart card data is that the information recorded only covers the stop-to-stop decisions within the public transport system. No information regarding access and egress (from/to the real origins and destinations of the trip) is recorded. Therefore, no behavioural analysis can be made about the choice of public transport stops or modes (i.e., walk to a close bus stop or a far metro station). This could be particularly serious in cases where non-walk modes (such as bicycle, park-and-ride, kiss-and-ride or taxi) are chosen to access or egress the public transport network, as they constitute multi-modal legs of the entire trip.

## 3.2 Modelling the Choice-Set

A significant advantage of smart card data is that a great amount of traveller's actual route choices become available, especially on systems with large penetration rates. Modellers can benefit from this fact to validate and enrich their choice-set generation methods. The massive information from smart card data has recently led to heuristic choice-set generation methods, where the choice-set for a particular origin-destination pair includes only the routes chosen by all the travellers (Raveau et al. 2011). This way, there is no need to generate any non-chosen routes, as they are not considered when modelling.

These heuristic methods could be directly applied to aggregated choice models, where no non-chosen route will be a part of the choice-set. Arguably, if no traveller in the system chooses a given route over a long period of time, that route may not be a part of the consideration set (and therefore, its predicted flow will be zero, as observed). From a disaggregated (i.e., individual) choice perspective, the assumption that non-chosen routes are not considered might not have a strong behavioural support. A given commuter might choose the same route every day because it is the best for his/her necessities and preferences, not because there are no other alternative routes. Even more, there is still a need in the literature for choice-set validation methods using smart card data.

## 3.3 Modelling Travel Times and Fares

Traditionally the most important variables used to explain route choice behaviour are fare and the travel time. Users tend to look for the fastest and least expensive way of getting from their origin to their destination and these two variables are the main criterion to discard unattractive (i.e., slow or expensive) alternatives. Regarding fares, their obtaining depends on the charging scheme and is generally straightforward. An advantage of smart cards is that they can record different traveller types that might be subject to different charging schemes (e.g., elders, students and concession cards).

Regarding travel time, many components could be considered: in-vehicle time, waiting time at the origin station and all subsequent transfer stations and walking time when transferring. As mentioned in Section 3.1, access and egress times are usually not available and therefore ignored. Ideally, these different time components might be considered separately to address their different perception and importance in the traveller's decision-making process. Nevertheless, smart cards can only record the times of tapping-in and (sometimes) tapping-out and therefore offer a leg time. If the tap-in happens at the station (like in metro and BRT systems), then the waiting time would be included in that leg time, but if the tap-in happens at the vehicle (like in traditional bus systems), then the waiting time would not be part of that leg time. Either way, it would not be possible to separately find travel time and waiting time.

## 3.4 Modelling Transfers

Regarding the transferring experience, the traditional approach is to consider the total number of transfers of each alternative route; as the real transferring time is captured by the walking and waiting time variables, this variable solely captures the displeasure of having to transfer. This information is generally obtained directly from smart cards. To further understand the transferring valuation, it is convenient to differentiate between possible types of transfers (Raveau et al. 2011; Raveau et al. 2014), in terms of stations layout, infrastructure, available travel information and additional services. This information has to be externally collected.

A traditional limitation of smart card data is that transfers within some modes (mainly metro systems) are usually not recorded. This way, the definition of trip legs is modified to model traveller's decisions, combining the unidentified legs on that particular mode (Tan et al. 2015). This is illustrated on Figure 3, where the smart card data fails to record: (i) the access leg, as the first record is the tap-in on Bus Line 1, (ii) the transfer between Metro Lines A and B and (iii) the egress leg, as the last record is the tap-out on Bus Line 2. As mentioned in Section 3.1, it is only possible to model the stop-to-stop route choice. An additional simplification, by not distinguishing the different travel alternatives inside the metro network, needs to be made to model the trip in this case.

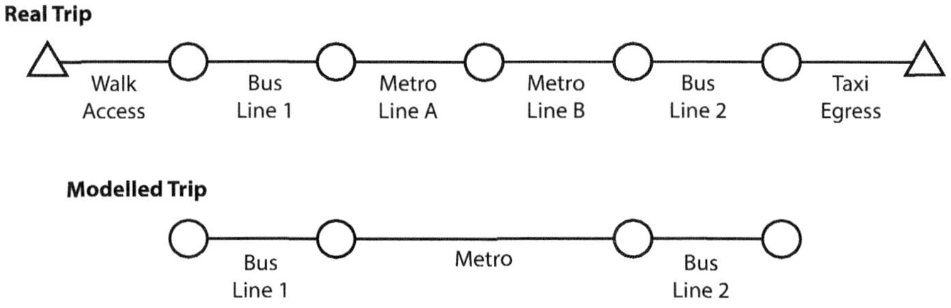


**Fig. 3.** Modelling unidentified legs

## 3.5 Modelling Comfort

The level of comfort and crowding experienced by the public transport users during their trip is also an important factor (Raveau et al. 2011; Tirachini et al. 2013). Capturing the comfort perception is not easy, as there is no clear measurement scale for comfort. One alternative is to use proxy variables, such as the mean occupancy along the route or the availability of air conditioning in the vehicles. In this sense, smart cards have become a significant source of information, as they have the potential to record the boarding and alighting of passengers for all vehicles at all stops. With this information, load profiles (and therefore crowdedness and occupancy indicators) could be obtained. Additional variables related to train usage, such as the possibility of getting a seat or the possibility of not boarding the first train could be considered. Regarding the possibility of not boarding, when that happens there is an excess waiting time that could be added to the time components mentioned above.

## 3.6 Modelling Individual Preferences

Traveller's socio-economic characteristics can influence their decisions and should be considered when modelling multi-modal route choices on public transport systems. When using smart card data, only some of the individuals' characteristics might be available. Among the ones that are usually not available are: (i) the purpose of the trip, (ii) the income level (especially relevant when related to fare), (iii) the gender and (iv) the age of the traveller. On the other hand, smart cards can collect other relevant information: (i) the time of the day when the trip begin (particularly peak or off-peak periods), (ii) the fare type (which might be also used to infer age, as students and seniors can have discount passes) and (iii) the frequency of the journey (e.g., daily, weekly, monthly, first time).

## 3.7 Modelling Travel Strategies

In available literature, it is usual to assume that all travellers are capable of considering high-complexity strategies (which might require developed analytical capacities). Similarly, it is usually assumed that all travellers have perfect information about the levels-of-service of all available alternatives. As expected, those assumptions cannot be true for a considerable proportion of the travellers and there is not enough empiric evidence to support (or disclaim) them. In this sense, smart card data could be a significant source of information, as it is possible to observe the repeated choices of individuals over long periods of time (mainly the high frequency trips, like work/school commute).

Based on the observed decisions for the same trip, it is possible to infer travel strategies: a given traveller might choose different bus lines between the same pair of stops on different days, which might be an indicator of common lines consideration (Schmöcker et al. 2013). The observed choice proportion of a given bus line can then be compared with the theoretical choice proportion according with the common lines theory, which depends on the bus lines frequencies. This way it is possible to get some insights about travel strategies.

# 4. CASE STUDY: SANTIAGO, CHILE

Multi-modal route choice models are applied to the public transport network of Santiago, Chile (6 million inhabitants). In Santiago, over 4 million trips are made daily on public transport modes. The public transport system (Transantiago) consists of 191 feeder (local) bus lines, 118 trunk bus lines and 5 metro lines. The demand is comprised of 730,605 trips in the morning peak period (6:30 AM to 8:30 AM). It is important to consider that, in Transantiago, travellers only tap-in when boarding the buses and accessing the metro system. Therefore, the alighting bus stops and metro stations has to be inferred (Munizaga and Palma 2012) to get the chosen routes. The smart card (demand information) was complemented with information on the levels-of-service (supply information) provided by the public transport authorities. The network is modelled with 616 one-way bus lines and 10 one-way metro lines. These lines generate a network with 852,548 line segments (which can be grouped to generate 663,696 route segments when considering common lines). There are 11,113 bus stops and 108 metro stations (modelled through 216 directional stops).

## 4.1 Choice-Set Generation

The set of alternative routes between public transport stops was generated using the link penalty approach (De la Barra et al. 1993). For each origin-destination pair, the shortest path was found using link-additive generalized costs, defined as a weighted sum of different attributes: fare, in-vehicle time, waiting time, walking time (when transferring), number of transfers (distinguishing between bus-to-bus, bus-to/from-metro and metro-to-metro), the possibility of travel seated and the possibility of not being able to board the first bus/train due to crowding. The weights for all these attributes were defined based on Raveau and Muñoz (2014).

The shortest path for each origin-destination pair in individual iteration is penalized, increasing its cost by 50%. With the updated costs, a new shortest path is found. This new shortest path is compared with the one(s) found earlier and kept in the choice set if the overlapping is less than 50% (in terms of shared route segments) with them. This process is conducted until three shortest paths that satisfy the overlapping criterion are found for each origin-destination pair.

The coverage of the choice set generation approach is 91%, as 663.599 of the 730.605 observed paths are recovered by the algorithm. This value is high, considering that only three alternatives are generated for each origin-destination approach. As the paths that travellers follow within the metro system are not recorded, the coverage of metro legs only takes into account access and egress stations (and not the potential paths that travellers may take within the system).

## 4.2 Model Specification

With the demand data obtained from the smart cards, route choice models can be estimated. For this, it is necessary to characterize the representative utilities of each available alternative. The attributes considered and the way they are obtained is described below:

***Fare***: Given the fare scheme in Transantiago, the fare of each alternative route depends exclusively on the usage of metro in any of the trip legs. Using metro has an additional cost of US$0.16 in the morning peak period.

***In-vehicle time***: Based on GPS data provided by the Santiago public transport authorities, bus travel times were calculated for the different road network links. Metro travel times were obtained from operational time tables, assuming the fastest path between access and egress metro stations.

***Waiting time***: Based on GPS data and recorded bus headway, empirical frequency distributions were found for each bus lines (these frequencies can differ from the operating plans due to congestion and bus-bunching). Following Welding (1957), expected waiting times were obtained for each bus stop in the network. The waiting times for metro lines correspond to half of their headway, due to its regularity, assuming minimum number of transfers (as each transfer means additional wait) within the metro network.

***Walking time***: As the smart card data covers stop-to-stop trips, the walking time corresponds only to transfers (access and egress are not captured). Transfer times within the metro system were obtained from field measurements (Raveau et al. 2011), assuming minimum number of transfers. The transfer times that involve bus were computed assuming a Manhattan walking grid between stops and a walking speed of 1 m/s.

***Transfers***: The number of transfers distinguishes four different transfer types: metro-to-metro, metro-to-bus, bus-to-metro and bus-to-bus. As the smart card data does not record the path decisions within the metro network, the metro-to-metro transfers correspond to the minimum number of necessary transfers between access and egress stations.

***Occupancy***: From the smart card data it is possible to get fairly accurate load profiles for the different bus lines. For the metro lines, trips within the metro systems were assigned using an existing route choice model defined specially for that mode (Raveau et al. 2011). The occupancy variable is defined as the distance-weighted ratio between passengers load and vehicle capacity. By definition the rate can vary between 0 (vehicles travelling empty along the entire route) and 1 (vehicles travelling fully loaded along the entire route).

***Possibility of seating***: This variable is related to the use of vehicle at low crowding levels distinguishing those stops where there is a possibility of getting a seat (depending on the occupancy of the vehicles when they leave the stop). In Transantiago this happens when the occupancy is 15% or less (these percentages represent the percentage of the capacity that corresponds to seats).

***Possibility of not boarding***: This variable is related to the use of vehicle at high crowding levels where there is a possibility of not boarding the first vehicle (and thus have to wait for the next vehicle). In Transantiago this happens when the occupancy is 85% or more.

***Angular cost***: To deal with the topology's effect on the route choices of the travellers, the model includes an angular cost to measure how direct a certain route is. Accordingly with Raveau et al. (2011) the angular cost is defined as shown by Equation (4.1), where $s$ represents a leg of the route, $d_s$ is the distance of leg $s$ and $\theta_s$ is the angle formed between the destination stop, the first stop of leg $s$ and the last stop of leg $s$.

$$\text{Angular Cost} = \sum_s d_s \cdot \sin\left(\frac{\theta_s}{2}\right) \tag{4.1}$$

***Commonality factor***: Finally, a commonality factor is incorporated to deal with the correlation between routes due to overlapping (Cascetta et al. 1996). This commonality factor is defined according to Equation (4.2), where $L_i$ is the total length of route $i$, $L_j$ is the total length of route $j$, $L_{ij}$ is the common length of routes $i$ and $j$ (due to overlapping), $\gamma$ is a positive parameter to be estimated and $A(q)$ are the available alternatives for individual $q$.

$$CF_i = \ln \sum_{j \in A(q)} \left(\frac{L_{ij}}{\sqrt{L_i \cdot L_j}}\right)^{\gamma} \tag{4.2}$$

## 4.3 Estimation Results

Based on smart card data from Transantiago, a C-Logit model (Cascetta et al. 1996; Cascetta et al. 2002) was estimated to understand the traveller's decision-making process while selecting public transport routes. The estimated parameters their t-values and goodness-of-fit indicators for the model are given in Table 1. It can be seen that all variables have the expected sign (with the exception of the possibility of seating, all level-of-service parameters are negative, as they represent a disutility) and are statistically significant at 95% confidence.

Based on the obtained parameters, it is possible to calculate monetary and temporal valuations for the different attributes (Table 2). These values correspond to the marginal rates of substitution with respect to the fare and the in-vehicle time. The angular cost and the commonality factor are excluded from this analysis, as they do not have a measurement scale. It can be seen than public travellers in Santiago value differently the different time components; among which the highest disutility comes from walking (due to the physical effort) followed by waiting (due to uncertainty). Amongst the transfer types, the worst kind is bus-to-bus (travellers are willing to travel up to 23 extra minutes to avoid them) while the least unpleasant is metro-to-metro. Bus-to-metro and metro-to-bus transfers are perceived similarly. The variables related to occupancy are significant, with travellers willing to travel 11 more minutes to find a seat or 16 more minutes to avoid a denied boarding.

**Table 1.** Estimation results

| Attribute | Parameter | t-value |
|---|---|---|
| Fare (US$) | −4.60 | −2.4 |
| In-vehicle time (min) | −0.11 | −9.4 |
| Waiting time (min) | −0.24 | −6.4 |
| Walking time (min) | −0.32 | −2.3 |
| Number of Bus-to-Bus transfers | −2.62 | −10.6 |
| Number of Bus-to-Metro transfers | −1.61 | −5.3 |
| Number of Metro-to-Bus transfers | −1.47 | −5.9 |
| Number of Metro-to-Metro transfers | −0.97 | −2.2 |
| Occupancy | −2.75 | −2.1 |
| Possibility of seating | 1.23 | 3.4 |
| Possibility of not boarding | −1.79 | −6.5 |
| Angular cost | −1.45 | −2.9 |
| Commonality factor | −0.76 | −3.0 |
| Sample size | 663,599 | |
| Log-likelihood | −668,814 | |
| Corrected $\rho^2$ | 0.386 | |

**Table 2.** Attribute valuations

| Attribute | Monetary Valuation | Temporal Valuation |
|---|---|---|
| 1 hour of In-vehicle time | 1.46 US$ | - |
| 1 hour of waiting time | 3.11 US$ | 2.13 hours In-Vehicle |
| 1 hour of walking time | 4.11 US$ | 2.81 hours In-Vehicle |
| 1 Bus-to-Bus transfer | 0.57 US$ | 23.41 minutes In-Vehicle |
| 1 Bus-to-Metro transfer | 0.35 US$ | 14.38 minutes In-Vehicle |
| 1 Metro-to-Bus transfer | 0.32 US$ | 13.14 minutes In-Vehicle |
| 1 Metro-to-Metro transfer | 0.21 US$ | 8.83 minutes In-Vehicle |
| 1% of occupancy | 0.60 US¢ | 0.25 minutes In-Vehicle |
| Possibility of seating | 0.27 US$ (1) | 10.98 minutes In-Vehicle[1] |
| Possibility of not boarding | 0.39 US$ | 15.98 minutes In-Vehicle |

[1] Absolute value, as this attribute represents a gain in utility.

## 5. CONCLUSION

Understanding public transport traveller's preferences and decision-making processes is essential in transportation planning to correctly predict travel decisions and the resulting flows on public transport networks. For this, it is necessary to identify the relevant factors which are considered and quantify the impact that different characteristics of the system have on their decisions. For this purpose, smart card data have arisen as a valuable information source in the past decade, as they can provide a significant amount of information related to the actual decisions of travellers. This information can be used to model their decision-making process and preferences through mathematical models.

Route choice modelling variables are traditionally limited to some tangible factors such as time and fare that, although relevant, fail to accommodate different aspects of traveller's behaviour. This chapter specifies and estimates a route choice model for the public transport system of Santiago, considering different types of variables: travel time components, transfers, occupancy and comfort indicators, network topology and path overlapping. All these variables are significant for understanding traveller's behaviour. This reassures the idea that public transport users take into account a variety of attributes when choosing routes.

Finally, from a social planning point of view, it has been shown and confirmed that travellers take into account a variety of attributes while choosing their routes and that their preferences can vary depending on their gender, the time of the day or the purpose of the trip. Travellers do not only care about travel times and number of transfers, but also care about crowding and topological factors. These results might be considered by the authorities and the planners, as many of the factors included in this study are not generally included in traditional route choice models.

## ACKNOWLEDGEMENTS

This research was supported by FONDEF D10I1049 "*Una herramienta táctico-estratégica de gestión y planificación de sistemas de transporte público urbano*". The author gratefully acknowledges the research support provided by the Centre for Sustainable Urban Development (CEDEUS) and the Across Latitudes and Cultures - Bus Rapid Transit (ALC-BRT) Centre of Excellence funded by the Volvo Research and Educational Foundations (VREF).

## REFERENCES

Bovy, P.H.L. 2009. On modelling route choice sets in transportation networks: a synthesis. *Transport Reviews* 29, pp. 43-68.

Bovy, P.H.L. and Hoogendoorn-Lanser, S. 2005. Modelling route choice behaviour in multi-modal transport networks. *Transportation* 32, pp. 341-368.

Bovy, P.H.L. and Stern, E. 1990. *Route Choice. Wayfinding in Transport Networks*. Kluwer Academic Publishers, Norwell, MA.

Cascetta, E., Nuzzolo, A., Russo, F. and Vitetta, A. 1996. A modified logit route choice model overcoming path overlapping problems. Specifications and some calibrations results for interurban network. *Proceedings of ISTTT Conference*, Lyon, France.

Cascetta, E., Russo, F., Viola, F.A. and Vitetta, A. 2002. A model of route perception in urban road networks. *Transportation Research Part B* 36, pp. 577-592.

Chriqui, C. and Robillard, P. 1975. Common bus lines. *Transportation Science* 9, pp. 115-121.

Daganzo, C. F. and Sheffi, Y. 1977. On stochastic models of traffic assignment. *Transportation Science* 11, pp. 253-274.

De la Barra, T., Pérez, B. and Anez, J. 1993. Multidimensional path search and assignment. *Proceedings of the 21st PTRC Summer Annual Meeting*, Manchester, England, pp. 307-319.

Dial, R.B. 1971. A probabilistic multipath traffic assignment model which obviates path enumeration. *Transportation Research* 5, pp. 83-113.

Hunt, J.D. 1990. A logit model of public transport route choice. *ITE Journal* 60, pp. 26-30.

McFadden, D. 1974. Conditional logit analysis of qualitative choice behaviour. In Zarembka, P. (ed.), *Frontiers of Econometrics*. Academic Press, New York, pp. 105-142.

Munizaga, M. and Palma, C. 2012. Estimation of a disaggregate multimodal public transport origin-destination matrix from passive smartcard data from Santiago, Chile. *Transportation Research Part C* 24, pp. 9-18.

Ortúzar, J. de D. and Willumsen, L.G. 2011. *Modelling Transport*. 4th Edition, John Wiley and Sons, Chichester.

Papinski, D., Scott, D. M. and Goherty, S. T. (2009). Exploring the route choice decision-making process: A comparison of planned and observed routes obtained using person-based GPS. *Transportation Research Part F* 12, pp. 347-358.

Prashker, J.N. and Bekhor, S. 2004. Route choice models used in the Stochastic user equilibrium problem: a review. *Transport Reviews* 24, pp. 437-463.

Prato, C.G. 2009. Route choice modelling: past, present and future research directions. *Journal of Choice Modelling* 2, pp. 65-100.

Ramming, M.S. 2001. *Network Knowledge and Route Choice*. Unpublished Ph.D. Thesis, Massachusetts Institute of Technology, Cambridge, MA.

Raveau, S., Guo, Z., Muñoz, J.C. and Wilson, N.H.M. (2014). A behavioural comparison of route choice on metro networks: Time, transfers, crowding, topology and socio-demographics. *Transportation Research Part A* 66, pp. 185-195.

Raveau, S. and Muñoz, J.C. 2014. Analysing route choice strategies on networks. *93rd Annual Meeting of the Transportation Research Board*, Washington, D.C., January.

Raveau, S., Muñoz, J.C. and de Grange, L. 2011. A topological route choice model for metro. *Transportation Research Part A* 45, pp. 138-147.

Schmöcker, J.D., Shimamoto, H. and Kurauchi, F. 2013. Generation and calibration of transit hyperpaths. *Transportation Research Part C* 36, pp. 406-418.

Spiess, H. and Florian, M. 1989. Optimal strategies: a new assignment model for transit networks. *Transportation Research Part B* 23, pp. 83-102.

Tan, R., Adnan, M., Lee, D.-H. and Ben-Akiva, M.E. 2015. A new path size formulation in path size logit for route choice modeling in public transport networks. *94th Annual Meeting of the Transportation Research Board*. Washington D.C., January.

Tirachini, A., Hensher, D.A. and Rose, J.M. 2013. Crowding in public transport systems: effects on users, operation and implications for the estimation of demand. *Transportation Research Part A* 53, pp. 36-52.

Welding, P.I. 1957. The instability of a close-interval service. *Operational Research Quarterly* 8, pp. 133-142.

## AUTHOR BIOGRAPHY

**Sebastián Raveau** is a Postdoctoral Associate at Singapore-MIT Alliance for Research and Technology (SMART), working on the Future Urban Mobility Interdisciplinary Research Group. He received his PhD from Pontificia Universidad Católica de Chile, where he is also an Associate Lecturer. His research interests focus on travel behaviour, travel demand analysis and transport systems. He is a keen collector of public transport maps and smart cards from across the world.

# PART 2

# Combining Smart Card Data with other Databases

Chapter 5

# Combination of Smart Card Data with Person Trip Survey Data

**T. Kusakabe[1,*] and Y. Asakura[2]**

**ABSTRACT**

The features of smart card data, i.e., precision, continuity and long-term observation enabled us to analyse the dynamic characteristics of travel behaviour. In order to explore the dynamic characteristics of a large amount of information in the data set, previous studies have developed data mining methodologies and applied them. However, the smart card systems were not specialized to collect a data set for travel behavioural analysis. The smart card data offer only fragmentary information on travel behaviour though they can provide accurate and continuous long-term data, which is difficult to achieve via conventional behavioural surveys. In order to supplement absent behavioural attributes in the smart card data, this study proposes a data fusion method of smart card data with the person trip survey data. The results of the data fusion enable us to analyse the continuous long-term features of the trip purpose of transport users, which are difficult to get from either the survey-based data or the smart card data. These results enable us to know specific behavioural segments, which caused changes in travel demand.

## 1. INTRODUCTION

Smart card systems have been widely installed as a method to collect fare of public transport. These systems automatically and continuously collect the records of passenger's use of the public transport with identification information. The amount of information storing in the data set is constantly increasing. And the data set is expected to include much

[1] Center for Spatial Information Science, the University of Tokyo, 5-1-5 Kashiwanoha, Kashiwa-shi, Chiba, 277-8568, Japan. Email: t.kusakabe@csis.u-tokyo.ac.jp

[2] Transport Studies Unit, Tokyo Institute of Technology, 2-12-1-M1-20, O-okayama, Meguro, Tokyo, 152-8552, Japan. Email: asakura@plan.cv.titech.ac.jp

* Corresponding author

traveller's information utilizable for transit planning, management and operation (see Pelletier et al. 2011). For example, as a management strategy, Asakura et al. (2008a) proposed a collaborative travel demand management (TDM) scheme in which a shopping centre provided incentives for customers who use railway transport. A railway company's smart card system used to verify the use of railway transport near the shopping centre. Asakura et al. (2012) showed an improvement in passenger behaviour caused by introduction of a new train timetable, which decreased the travel time between several stations. They showed analyses of the departure, travel and arrival time distributions using smart card data. Thus, precise, continuous and long-term observation data could be used to show the dynamic characteristics of travel behaviour even though the records in the data set only include information on the date, time, place of boarding/alighting and ID.

## 1.1 Exploration of Smart Card Data Set Using Visualization

Visualization is one of the fundamental methods used to explore the dynamic characteristics of a large amount of information in a data set of smart cards. Figure 1 shows an example of the visualization results, which describe the potential of the smart card data set for understanding travel behaviour. The figure shows the amount of passengers who boarded the trains, according to the date and time at Station I. Location of Station I is in the central business district (CBD) in the Osaka metropolitan area in Japan. The horizontal axis indicates the time of day and the vertical axis indicates the date. The grey scale represents the number of trips. Although this visualization method does not consider ID information, several characteristics of passenger behaviour are distinguishable. For example, vertical stripe is found in the morning peak but they are not found in the evening peak. This example shows that passengers travel according to the train schedule in the morning and morning commuters are more sensitive to time than evening passengers.

Asakura et al. (2008b) proposed another visualization method for smart card data recorded at the ticket gates of both the origin and destination stations. Figure 2 shows a relational map of the scheduled train times and gate passage times derived from the transaction data. The horizontal axis indicates the time of arrival and the vertical axis indicates the time of departure. The dots show the time when the passengers pass through the gates at the departure and arrival stations. Each record from the transaction data is plotted as a dot in this map. The horizontal lines on the map represent the departure times of trains and the vertical lines represent the arrival times. Passengers can board trains that are plotted above the dot and they can alight from trains plotted on the left side of the dot. A passenger can take any train that departs after their entry at the departure station only if that train arrives before the passenger's exit time at the arrival station. As shown in the figure, the passenger can choose from several train types, which have different stops and travel times. The passengers tended to use faster train services. However, a few passengers still used slower services to avoid congestion.

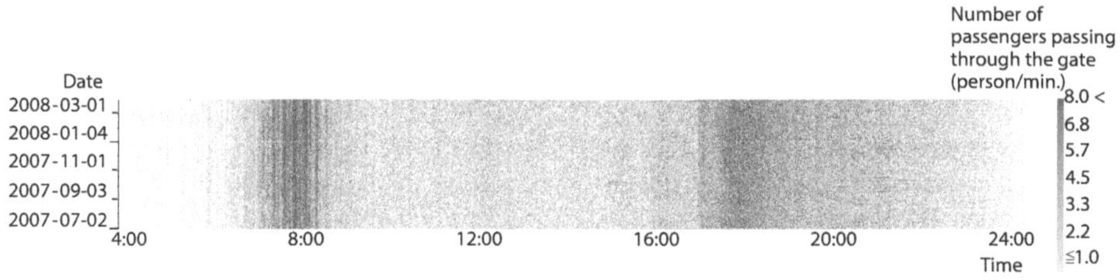


**Fig. 1.** Simple visualization of smart card data – number of passengers who boarded at Station I

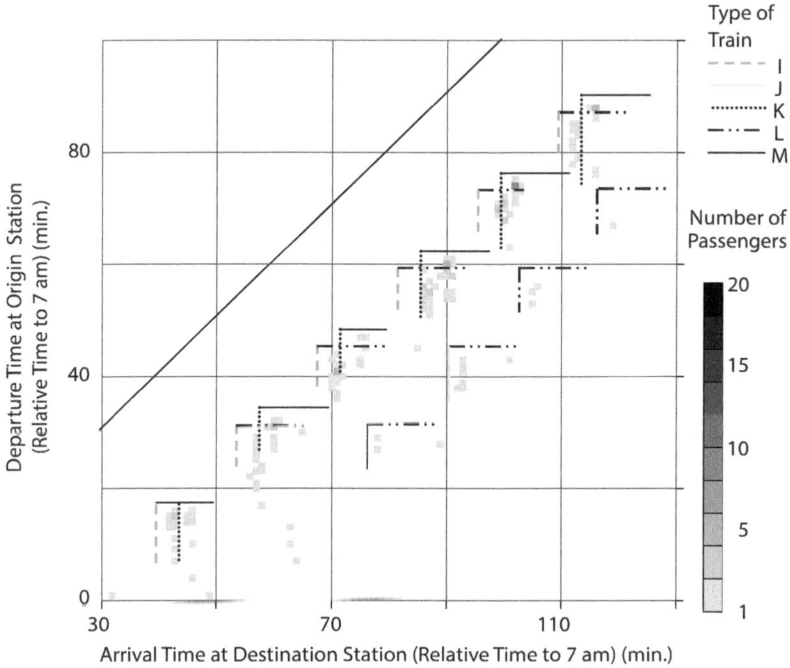


**Fig. 2.** Relational map with actual smart card data and train timetable (Kusakabe et al. 2010)

## 1.2 Interpretation of Features of Smart Card Data Set

If an analyst has experience of and knowledge about a target transport system, he/she can intuitively and easily determine characteristics of data using the simple visualization methods owing to the precise and continuous data set. However, the certainty of interpretation of the characteristics depends on the ability of the analysts. This is because the data set is fragmentary for behavioural analysis. For example, the data do not directly include the passenger's origins, destinations, or trip purposes and the data do not represent travel behaviour throughout the entire transport network.

In order to achieve a consistent interpretation of the data set, several studies have attempted to develop methods to estimate the user segments or the behavioural contexts from behavioural patterns observed using smart card data. Kusakabe et al. (2010) developed an algorithm to estimate the most likely train boarded by a passenger. Their method relied on train timetable data and precise time records obtained at both the boarding and alighting gates. By using the method, the relationship between the train choice and the timetable were analysed.

Passenger ID is also useful information to analyse travel patterns. It enables us to analyse each passenger's trip frequency, travel sections and trip sequences. Travel patterns and their variability over long-term periods can thus be analysed (e.g., Bagchi and White 2005 and Utsunomiya et al. 2006). Agard et al. (2006) and Morency et al. (2007) determined behavioural pattern groups and showed the variability in traveller behavioural patterns. Kusakabe and Asakura (2011) proposed a method to classify within-day and day-to-day behavioural patterns of smart card users using a latent class model. However, the meaning of the segments of the smart card users, determined by the behavioural patterns, should be subjectively interpreted by analysts in these studies.

## 1.3 Interpretation of Features Using Data Fusion

The smart card data offer only fragmentary information on travel behaviour though they can give continuous long-term data, which is difficult to achieve via conventional behavioural surveys. For example, the possible duration of travel-behaviour surveys in the previous studies is less than a few months even if the survey employs technologies such as a global positioning system (GPS), which enables us to automatically track respondents (e.g., Murakami and Wagner 1999; Asakura and Hato 2004; Wolf et al. 2001 and Draijer et al. 2000). In contrast to smart card data, survey-based data could be used to directly get detailed information on travel behaviour. If these data are integrated, there could be understanding of the relationship of behavioural attributes that cannot be obtained from either smart card data or survey-based data alone. Data fusion is one of the approaches to integrate multiple data sources and it is applied in various fields, such as the military, marketing and intelligent transportation systems (e.g., Hall 1992; Mitchell 2007; Kamakura and Wedel 1997; El Faouzi et al. 2011, Shen and Stopher 2013; Kusakabe and Asakura 2014 and Gong et al. 2014). For example, Shen and Stopher (2013) developed a trip purpose imputation method for GPS data by using the National Household Travel Survey (NHTS) in the US. In their method, the trip purpose, which was not directly, determined using GPS data, was estimated using the rules obtained from the NHTS data.

In this chapter, application of the data fusion method proposed by Kusakabe and Asakura (2014), to estimate the trip purpose of passengers from smart card data gained at several stations. The proposed method is to enhance the understanding of travel behaviour during monitoring of the smart card data. The results of the data fusion enable us to analyse the continuous long-term features of the trip purpose of transport users, which are difficult to get from either the survey-based data or the smart card data. The analysis in this study focuses on finding the relationship between the estimated trip purpose and travel patterns. The expected result is to suggest specific behavioural segments, which cause changes in travel demand.

Section 2 presents the proposed data fusion method for smart card data and person trip survey data. Section 3 presents the empirical analysis and validation to confirm that the method can show changes in behavioural features of transport usage observed using the smart card data. Section 4 concludes this study.

## 2. MODEL

This study employs the method proposed by Kusakabe and Asakura (2014), based on the Naïve Bayes classifier (Rish 2001). Their proposed model estimated the trip purposes of transit smart card users by combining the information from person trip survey data. The model was calibrated using the survey data and then applied to smart card data to estimate the trip purposes. Their proposed framework could be used for other travel context estimations such as origin and destination estimations for a trip.

Section 2.1 shows the data structures discussed in this study. Section 2.2 presents an overview of the proposed data fusion method. Section 2.3 presents the method using the Naïve Bayes probabilistic model. Section 2.4 describes the estimation of the Naïve Bayes probability functions.

### 2.1 Schema of Smart Card Data and Person Trip Survey Data

Records of trips using railway transport in the person trip survey data contain the ID of a card, trip ID, origin and destination of the trip, departure and arrival times and trip purpose. The boarding and alighting stations and times are also included. The data items in the smart card data include the card ID, date, boarding station/time and alighting station/time. The two data sets do not have common IDs. Although these two data sets are collected separately, both the smart card data and the person trip survey data contain information on the boarding and alighting stations and times.

The information on the boarding and alighting stations and time could be used to combine the two data sets. However, these attributes are not exactly same for the following two reasons. First, the accuracy of the information is different. The smart card data contain the exact minute when a traveller passes through the gate at the station. However, the boarding and alighting time in the person trip survey data is reported after the trip. The information in the person trip survey is sometimes not correct and travellers possibly report the time rounded by 5 or 10 minutes because they report it from memory. This affects the time resolution of the discrete variables used in the proposed method. Second, the IDs of the smart card data are not always identical to the people. For example, a smart card could be shared among family members or a traveller can have more than one card. However, such cases cannot be common especially for registered monthly passes in Japan that associate a specific user with a smart card.

## 2.2 An Overview of Data Fusion Method

In previous studies, many types of machine learning methodologies have been implemented to estimate the purpose of trips in a passive data set such as GPS data (see Gong et al. 2014). One of the advantages of the Naïve Bayes classifier is that it requires a smaller amount of training data compared to other machine learning methods. The model is further described by using the simple probability functions that could be easily understood by analysts. However, this method does not deal with many correlated variables. This is because the Naïve Bayes classifier assumes that each element of an explanatory variable is conditionally independent of every other element to reduce the required number of data. When larger amount of attributes and larger training data sets are available compared with those of the person trip survey data, the estimation method could be replaced with a machine learning method more advanced than the Naïve Bayes classifier.

Figure 3 shows the flow chart of the proposed data fusion method. The concept of data fusion is to estimate absent attributes of respective data sets. The behavioural attribute $c$ represents the attribute observed only in the person trip survey data, such as trip purpose, origin and destination. Kusakabe and Asakura (2014) employed trip purpose as the behavioural attribute $c$. Definition of the possible values of $c$ is:

C = {'commuting to work or school', 'leisure-or-business', 'returning home'}. (2.1)

Note that business purposes show the trips where travellers travel between their workplace and other places except their homes such as client offices.

The attribute $F$ represents the commonly observed behavioural attributes included in both the data sets, such as boarding stations and times. Specifically, Kusakabe and Asakura (2014) employed $F = \{f_a, f_s\}$ where $f_a$ is 'alighting time' and $f_s$ is 'duration of stay'. The $f_a$ represents the time of the trip. Definition of the $f_s$ is the interval between the alighting time and the next boarding time at the same station. Both are discrete variables defined every hour. The duration of stay implicitly represents the duration of stay at the destination in addition to the travel time between

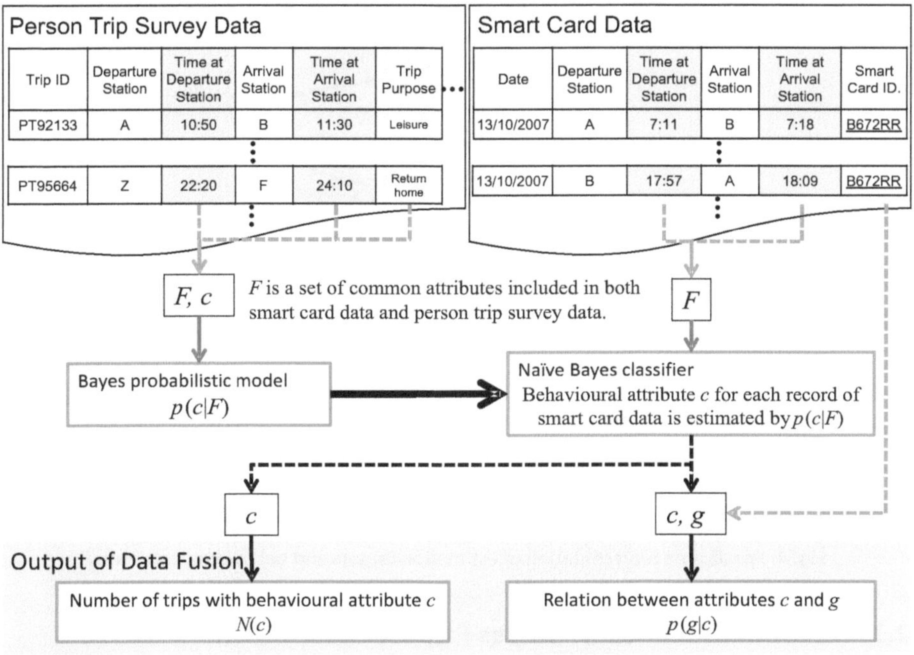


**Fig. 3.** An overview of data fusion method

the alighting station and the real destination. The travel time includes the access and egress times with different modes of transport such as taxis and buses. The behavioural attribute *g* could only be derived from smart card data, such as trip frequency. A continuous collection method is required to derive *g*. The proposed method could be used to determine the number of trips with trip purpose *c* from the smart card data. In addition, the method provides the relationship between *c* and *g*, which cannot be obtained from only one data set.

The conditional probability distribution $p(c|F)$ represents the probability, where the trip purpose is *c* at the gate of the station for attribute *F*. The distribution is estimated from the person trip survey data. By applying $p(c|F)$ to the Naïve Bayes classifier, the trip purpose *c* is added to the data for each trip in the smart card data. This enables us to analyse time series changes in $N(c)$, the number of trips with trip purpose *c*. Additionally, the relationship between *c* and *g* could be summarized using $p(g|c)$, which is the conditional probability distribution of *g* for *c*.

Figure 4 shows a conceptual representation of the estimation target *c* and *F* in the space-time dimensions. In the figure, presentation of the trip for the estimation where a traveller alights at Station A is given by a bold line. The trip purpose *c* of the trip is estimated from $f_a$ and $f_s$ by using $p(c|F)$. Note that the ID information of a traveller is required to derive $f_s$ because the data of this variable are determined by two consecutive trips.

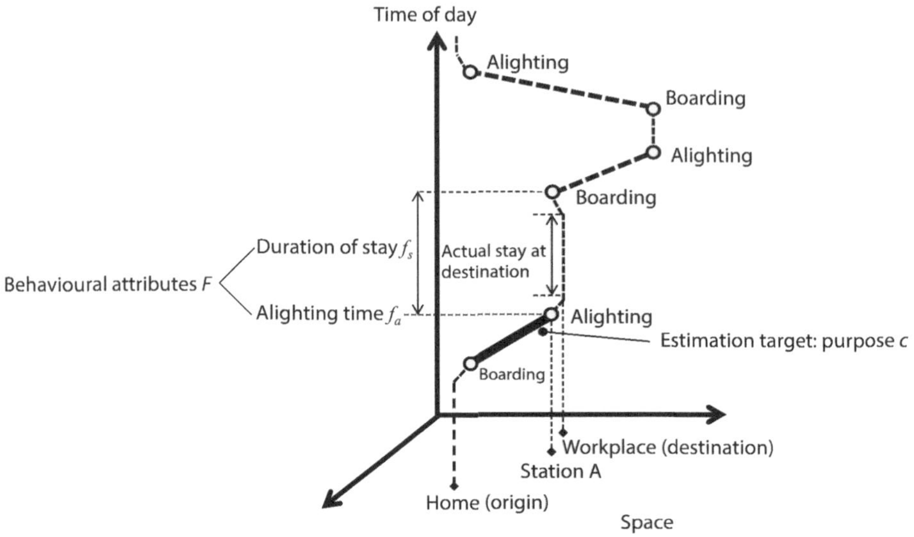


**Fig. 4.** Space-time representation of estimation targets and behavioural attribute *F*

## 2.3 Formulation of Naïve Bayes Probabilistic Model

By using Bayes' theorem, $p(c|F)$ could be expressed using,

$$p(c|F) = \frac{1}{p(F)} p(c) \prod_{k \in \{a,s\}} p(f_k|c) \tag{2.2}$$

where, $p(c)$, $p(F)$ and $p(f_k|c)$ are probability distributions estimated from the person trip survey data. The distributions $p(c)$ and $p(F)$ are derived from the composition rate of trips having attributes $c$ and $F$, respectively. The conditional probability distribution $p(f_k|c)$ is derived from the percentage of trips having attribute $f_k$ corresponding to each value of trip purpose $c$.

By using $F$ of each trip given by the smart card data, the trip purpose $c$ of each trip is estimated using the Naïve Bayes classifier. The equation for the classifier is as follows:

$$\hat{c}(F) = \arg\max_{c \in C} p(c|F) \tag{2.3}$$

Note that considering $p(F)$ as a constant because it does not depend on $c$. By using Equation (2.3), the number of trips with behavioural attribute $c$ is expressed as,

$$N(c) = \sum_{F \in S} \delta(c, F) N_s(F) \tag{2.4}$$

where,

$$\delta(c,F)=\begin{cases}1 \text{ if } \hat{c}(F)=c\\ 0 \text{ otherwise,}\end{cases}$$

$N_s(F)$ is the number of trips with a vector of attribute $F$ determined by using the smart card system and $S$ is a set of all the possible values of $F$.

Using a variable $g$ derived from the smart card data, distribution of the trip purpose $c$ of each $g$ could be estimated using Bayesian inference. The joint probability of trips whose attributes are $c$ and $g$ are given by,

$$p(c,g)=\sum_{F\in S} p(c|F)\,p_s(F,g) \tag{2.5}$$

where $p_s(F, g)$ is derived from the composition rate of trips having $\{F, g\}$, which is determined by using the smart card data. Then, the distribution of $g$ for each $c$ is calculated as the posterior distribution using the person trip survey and smart card data. This is described by,

$$p(g|c)=\frac{p(c,g)}{p(c)}=\frac{\sum_{F\in S} p(c|F)\,p_s(F,g)}{\sum_{F\in S} p(c|F)\,p_s(F)} \tag{2.6}$$

where, $P_s(F)$ is the composition rate of trips with a vector of attribute $F$, determined by using the smart card system.

## 2.4 Estimation of Probability Functions

Each conditional probability function is obtained from maximum likelihood estimation. The probability trip purpose $c \in C$ when a value of attribute, $f_k \in F_k$, $k \in \{a, s\}$, is given and expressed as $p(f_k \mid c) = p_{cfk} \in p_{kc}$. The probability distribution $p_{kc}$ is given by the solution of the following maximization problem:

$$\max_{P_{kc}} l(P_{kc})=\prod_{a\in A}\prod_{f_k\in F_k} p_{f_k c}^{\delta(a,c,f_k)} \tag{2.7}$$

$$\text{s.t.} \sum_{p_{f_k c}\in P_{kc}} p_{f_k c}=1 \tag{2.8}$$

$$\delta(a,c,f_k)=\begin{cases}1 & \text{if } c_a=c\cap f_{ka}=f_k\\ 0 & \text{otherwise}\end{cases} \tag{2.9}$$

where, $A$ is the data set of person trip survey data, $r = (c_a, f_{ka})$ is the data that is a subset of $A$ and $c_a$ and $f_{ka}$ are trip purpose and attributes. This maximization is converted to the following problem using a log-likelihood function and the method of Lagrange multipliers.

$$\max_{P_{kc},\lambda} L(P_{kc},\lambda \mid c,t) = \sum_{r\in A}\sum_{f_k\in F_k}\delta(r,c,f_k)\log(p_{f_kc}) - \lambda\left(\sum_{p_{f_kc}\in P_{kc}} p_{f_kc} - 1\right) \quad (2.10)$$

Hence, the solution of this maximization is,

$$P(f_k \mid c) = p_{cx_k} = \frac{\sum_{r\in A}\delta(r,c,f_k)}{\sum_{f'_k\in F_k}\sum_{r\in A}\delta(r,c,f'_k)} \quad (2.11)$$

The probability of trip purpose is also obtained in the same manner. The description of log-likelihood maximization is given by,

$$\max_{P_C,\lambda} L(P_C,\lambda \mid t) = \sum_{r\in A}\sum_{c\in C}\delta(r,c)\log(p_c) - \lambda\left(\sum_{c\in C} p_c - 1\right) \quad (2.12)$$

$$\delta(r,c) = \begin{cases} 1 & \text{if } c_a = c \\ 0 & \text{otherwise} \end{cases} \quad (2.13)$$

where, $p(c) = p_c$, $p_c \in P_c$ is probability of the trip purpose $c \in C$. The solution to this problem is given by,

$$P(c) = p_c = \frac{\sum_{r\in A}\delta(r,c)}{\sum_{c'\in C}\sum_{r\in A}\delta(r,c')} \quad (2.14)$$

## 3. EMPIRICAL ANALYSIS

This section describes an empirical analysis using the proposed method. The smart card data sets for two railway stations are employed. Section 3.1 discusses the data sets. Section 3.2 describes the validation analyses that use a subset of the person trip survey data. By comparing the estimated trip purpose with the real trip purpose obtained using the person trip survey, this section examines the accuracy of estimation. Section 3.3 applies the proposed method to the real smart card data to find long-term changes in traveller usage of stations.

## 3.1 Data Sets

This study employs the data obtained at Station A and B. These target railway stations for the analysis are operated by a private company and are in the Osaka area, the second largest metropolitan area in Japan. Some railway lines run parallel to each other; hence, travellers can choose their train from several railway lines. However, the data for other railway companies were not available for this analysis. The proposed method is applied to the smart card transaction data, obtained from only one railway operator.

The contents of the transaction data records are described in Section 2.1. Approximately, 10% of the passengers of the railway company were smart card holders. Since, the railway company allowed the use of the smart card data only for research purposes, the cards ID information was anonymized before the analysis. The privacy of the smart cardholders was strictly protected throughout this study.

Location of Station A is one of the major stations in the CBD in the Osaka area. The station has many railway connections to various districts, operated by more than one railway company. Location of Station B in a residential district near several schools. The data for estimating the probability distributions of trip purposes were person trip survey data obtained in 2002 during the '4th Kei-han-shin Metropolitan Area Person Trip Survey.' The data for the trips where passengers alight at each target station operated by the same railway company, used to estimate the model. The duration of stay was calculated using the alighting time and boarding time at each target station.

The person trip survey data for travellers who alighted at each target station contained records of 1,586 trips made by 1,576 travellers for Station A and 211 trips made by 208 travellers for Station B. The data set for each station was randomly divided into two subsets: estimation and validation data sets. The probability distribution $p(c|F)$ of Station A estimated from 1,095 trips and that for Station B estimated from 132 trips. The validation data consisted of 491 trips for Station A and 77 trips for Station B.

Section 3.3 discusses the genuine smart card data observed in 20 months from October 2007 to May 2009. All the person trip survey data for passengers alighting at the target stations were used to determine the models. This analysis used the smart card data of travellers who alighted at each target station at least once during the data collection period. The smart card data for Station A covered 7,074,768 trips made by 553,259 travellers and that for Station B covered 667,132 trips made by 69,204 travellers.

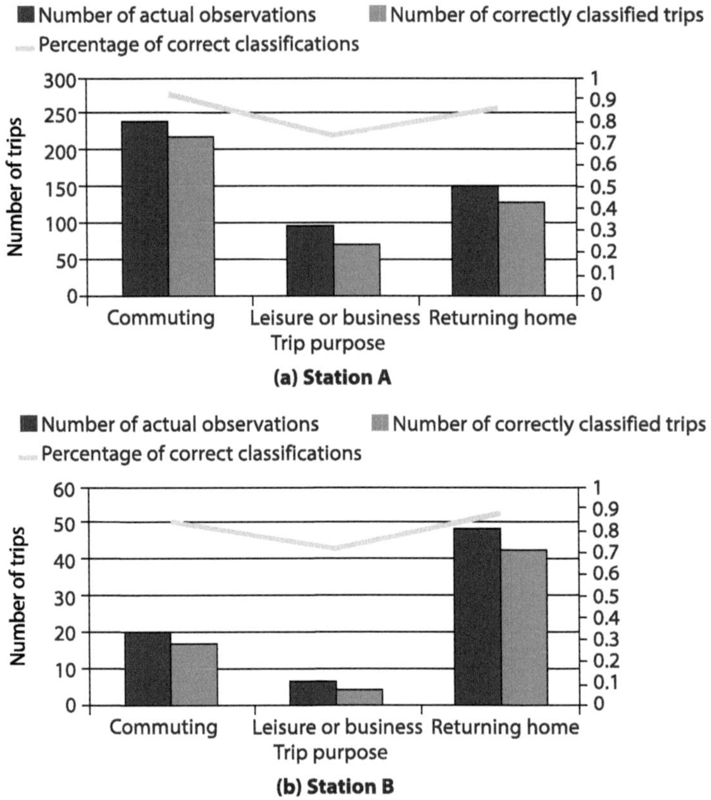


**Fig. 5.** Number of trips at the target station correctly estimated by trip purpose

## 3.2 Validation with Person Trip Survey Data

This section examines the model of the proposed method using the validation subset of the person trip survey data described in Section 3.1. The data include both attributes $c$ and $F$ and are observed in a day. To confirm whether the trip purpose was correctly estimated using Equation (2.3), the estimate trip purpose is compared with the real trip purpose.

Figure 5 shows the estimation results of the trip purpose using Equation (2.3) and Equation (2.4). The number of the real trips is the one appeared in the validation data. The number of estimated trips was determined using Equation (2.4). The number of successfully estimated trips indicates the trips for which the actual purpose was the same as the estimated purpose according to Equation (2.3). The trip purposes in the figure are defined in Equation (2.1). For Station A and B, 86.2% and 85.5% of the trips were correctly estimated. More than 80.0% of the commuting trips and returning-home trips were correctly estimated. Especially, the correctly estimated trips of the commuters for Station A were 92.1%. In contrast, relatively few leisure-or-business trips were correctly estimated. One of the possible reasons for this low accuracy is that these trips are few. The leisure-or-business trips observed in the estimation data set were no more than 26% of the total. For Station B, they were only 20.9%.

## 3.3 Application to Data Mining of Smart Card Data

This section describes application of the data fusion method to actual smart card data observed over 20 months. The purpose of the analyses was to find the characteristics of day-to-day changes in the behavioural features. First, it discusses the day-to-day changes in the number of trips for each trip purpose as estimated using Equation (2.4). We conclude this section by showing the month-to-month changes in the distribution of the trip frequency as derived from Equation (2.6). This analysis shows the relationships between the estimated trip purposes $c$ and trip frequency $g$ determined from the smart card data. These relationships can illustrate the characteristics of changes in the demand because there is effect on changes in the number of trips by the change in the number of travellers as well as the number of trips made by each traveller.

Figure 6 shows the day-to-day changes for Station A and B in the number of trips for each purpose as estimated using Equation (2.4). It describes the number of trips of each purpose, for each day for 20 months. The number of commuting travellers alighting at Station A was twice that of the "returning-home" or "leisure-or-business" travellers because location of Station A is in the CBD. The commuting trips accounted for 50.6% of the total, the leisure-or-business trips for 22.8% and the returning-home trips for 26.6%. On the other hand, for Station B, the returning-home travellers were more than twice the commuting travellers because location of this station is in the residential district. The shares of the commuting trips, leisure-or-business trips and the returning-home trips of Station B were 29.0%, 20.5% and 50.5% respectively.

Overall, the number of trips increased during the observation period. The average number of trips in October 2008 for stations A and B increased by 35.2% and 46.7% compared with those in October 2007. Although we cannot distinguish whether the changes were caused solely by variations in demand or in the composition rate of smart card holders, several different characteristics of changes were observed corresponding to trip purposes and stations. For example, the leisure-or-business trips for Station B significantly increased in March 2008. The number of leisure-or-business trips in October 2008 increased by 76.5% compared with that in October 2007. This percentage is larger than the average increase for other trip purposes. This was probably due to the opening of a new shopping mall near the station.

For Station A, large variations in the returning-home trips were observed in the summer. The large variations coincided with the days when baseball games were held. This might be due to travellers transferring to other railway lines on their way home from the stadium at other station in the target line. On 94% of the days when baseball games were held, the number of travellers between 8 p.m. and 9 p.m. increased by 20% compared to the average. On the other hand, the number of travellers on other days did not exceed 1.2 times of the average. By using the Welch's t test, the estimated t-value was 14.1; that is, the number of travellers on the days of the baseball games was larger than the usual. This result was among those used to find events from the data without presumption of the events. The ability to find the events without presumption will help to know the period and stations affected by the events before conducting the detailed surveys.

There was a large decline in the commutes during the summer and New Year holiday seasons for Station A; the number of trips was less than 60% of the average on 12 days. However, for Station B, a decline in the commutes during school holidays is found, which was larger decline than ordinary holidays. For Station B, the number of trips was less than 60% of average on 27 days.

Figures 7 and 8 show distribution of the trip frequency for commuting, leisure-or-business and returning-home trips, estimated using Equation (2.6). These results show time series changes in trip purposes, which are difficult to find from person trip survey data. The vertical axis shows the number of the days per month where each traveller made trips from the station. The horizontal axis shows the month. The grey scale indicates the composition rate for the number of days in each month when the travellers used the station. These figures show the month-to-month changes in the trip frequency for each trip purpose.

Most commuting travellers made their trips using Station A on all the weekdays. However, the frequency of commutes decreased in the holiday seasons, i.e., January, August and December. Most of the commuting travellers made trips in December 2007, January 2008 and August 2008 on 19, 18 and 18 days. These figures show the variability of the frequency tended to increase while the frequency itself decreased in these seasons. This could be confirmed using the composition rate on the day when the most commuting travellers made trips. The composition rates in December 2007, January 2008 and August 2008 were 20.2%, 19.6% and 12.9%, respectively, even though the average rate in months except January, August and December was 22.7%. For Station B, the frequency of commutes clearly decreased in the holiday seasons compared with that for Station A. The frequency of commutes seemed to decrease by school holidays. These results imply that the large decrease in the number of commutes during holiday seasons in Figure 6 caused by a decrease in the trip frequency. On the other hand, most leisure-or-business travellers were recorded once in a month at either station. This result implies that the changes in the number of the leisure-or-business trips were caused by changes in the number of travellers.

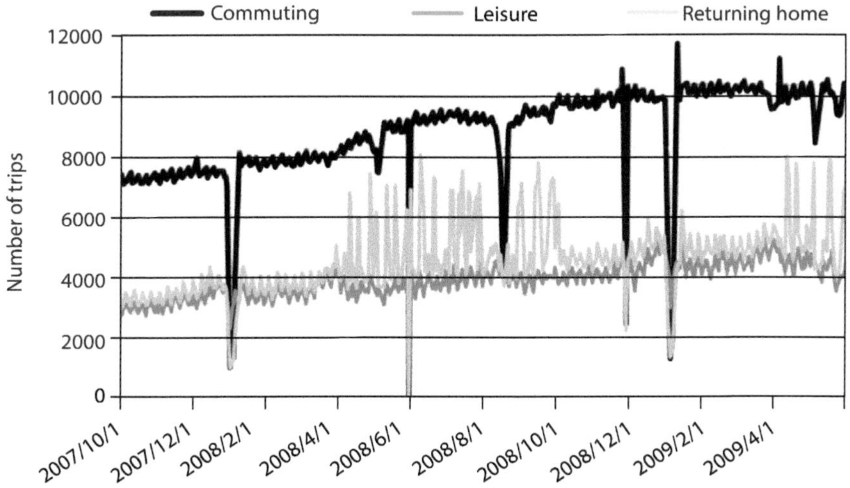


**(a) Station A**

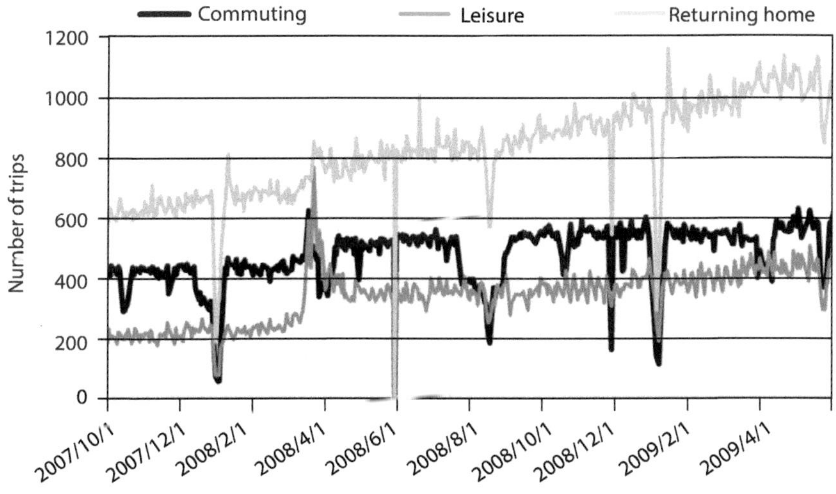


**(a) Station B**

**Fig. 6.** Day-to-day changes in the number of trips

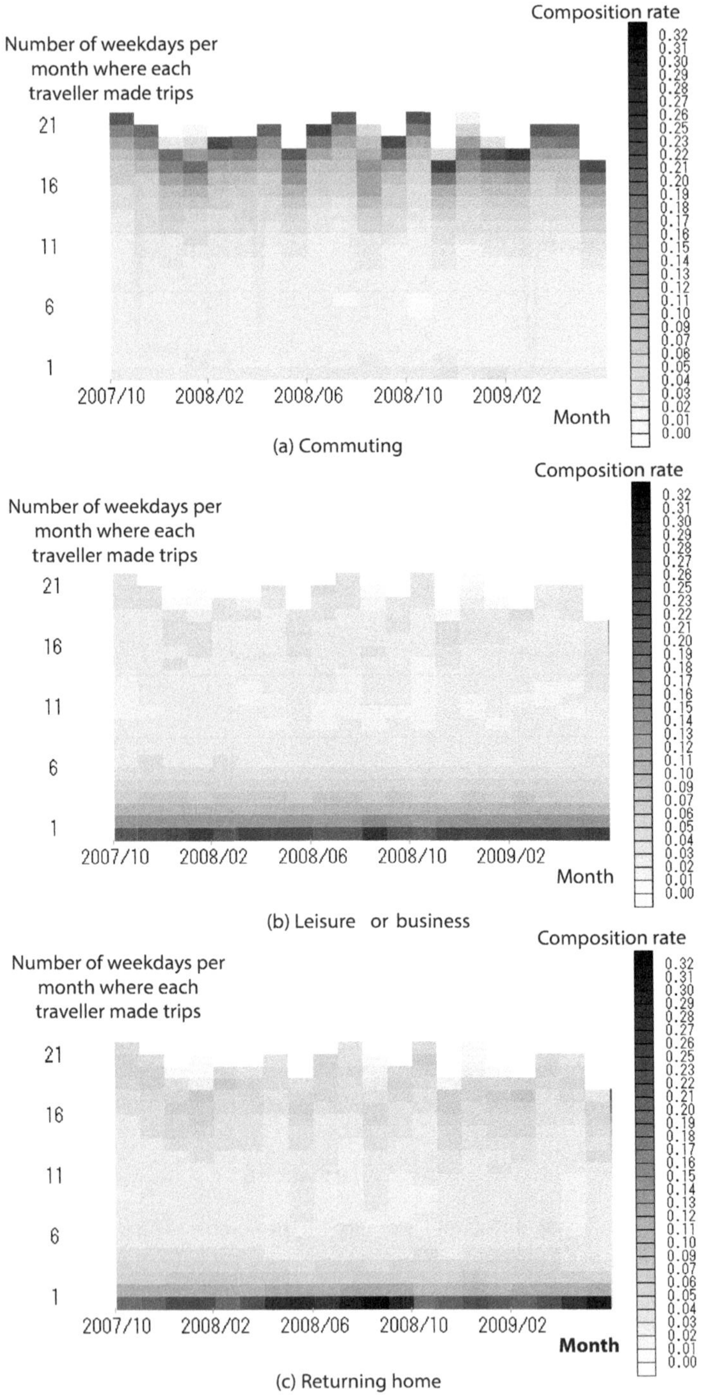


**Fig. 7.** Histograms of the number of weekdays where each traveller made trips in each month at Station A

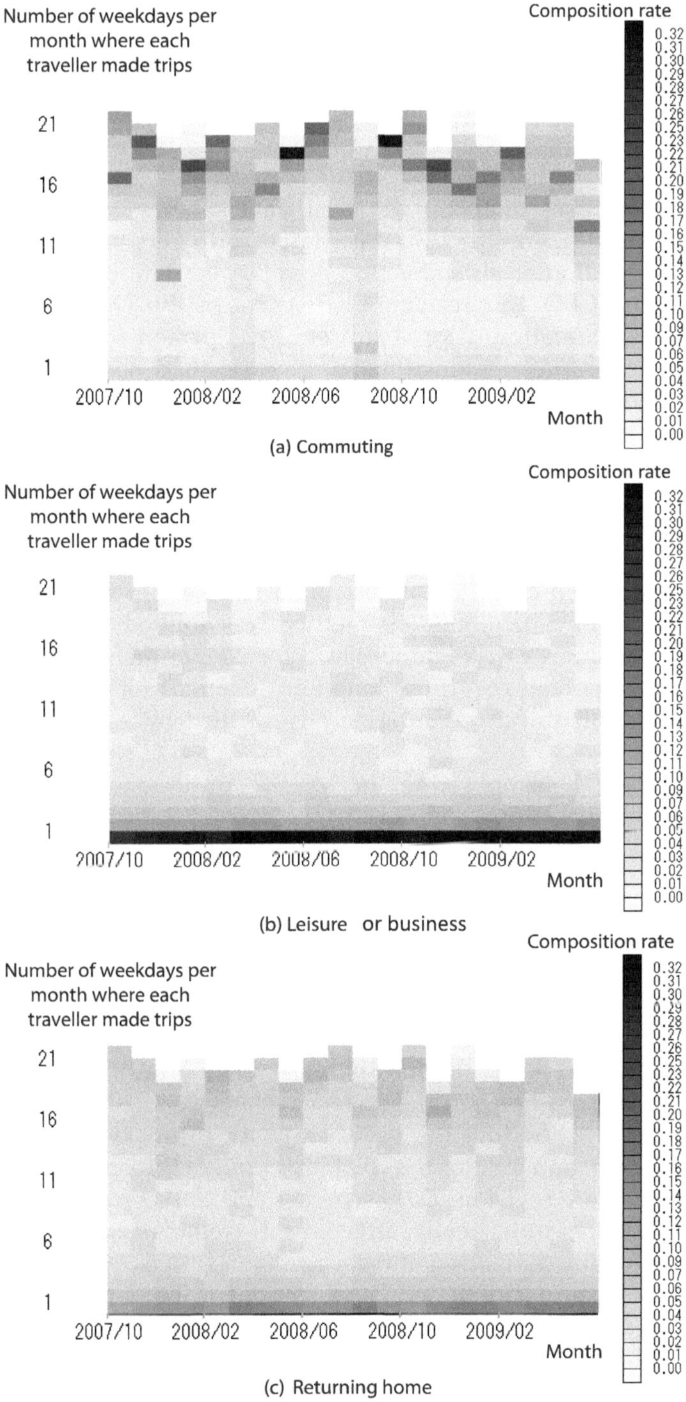


**Fig. 8.** Histograms of the number of weekdays where each traveller made trips in each month at Station B

## 4. CONCLUSION

The features of smart card data, i.e., precision, continuity and long-term observation enabled us to analyse the dynamic characteristics of travel behaviour. However, the smart card data include fragmental information for travel behaviour data. In order to quantitatively complement the behavioural features to the smart card data set, we used a data fusion method, which combines the smart card data with person trip survey data.

The empirical analysis in Section 3 employed data obtained from two stations in the CBD and residential districts of the Osaka metropolitan area in Japan. Validation using a subset of the person trip survey data, as shown in Section 3.2, demonstrated that more than 80.0% of commuting trips and returning-home trips were correctly estimated for any of the stations. The empirical data mining analysis using the real smart card data set in Section 3.3 showed that the proposed method was capable of helping us to find and interpret the behavioural features observed in the smart card data. The proposed method illustrated the share of trip purposes among travellers and the relationship between the trip frequency and the trip purpose, which could not be obtained from either smart card data or person trip survey data alone. The method was applied to the data mining analysis of the smart card data observed for 20 months, obtained from the two stations. The features in the long-term changes for each trip purpose could be illustrated and quantitatively confirmed by using the estimated trip purposes.

By applying the proposed method to the continuous monitoring of the passengers, transport operators can make assumptions about the cause of behavioural changes. For example, this study represented the user groups according to the trip purposes, which contributed to the change in demand. Continuous monitoring will help the operators to find the amount of effect as well as the spreading speed as a result of policy changes. It will also help to find suitable survey areas, targets and time periods for conducting the surveys. For example, as shown in the empirical analysis in this study, large variations were observed in the returning-home trips during the summer for Station A but not for Station B. Owing to this result, Station B could be excluded when a transit agency plans detailed surveys on these trips. This will also reduce the survey cost. In future work, the proposed method will be applied to the real assessment of specific operational improvements, fare revision, TDM schemes and transit planning to confirm whether these measures really affect the traveller's segment in the way expected before implementing of measures.

When the composition rate of a smart cardholder is low, the data is used to determine user segments that often use smart cards. If the composition rate becomes high enough, the data could be used for analysing the demand of public transport. Especially, when the dependency on public transport in an area is high, the data include more trip-chains made by public transport. Such data may enable us to analyse places of activity and changes in following years. The characteristics of activities might be analysed to socio-demographic factors obtained from other surveys, such as age composition.

Regarding methodological aspects, this study employed the Naïve Bayes classifier for which the required data is relatively less than that for other machine learning methodologies. The method was suitable for data fusion of smart card data with person trip survey data because of the limited data. If larger scale and richer behavioural data than person trip survey data, such as advanced tracking survey data (e.g., Asakura and Hato 2004; Cottrill et al. 2013; Kusakabe et al. 2015), becomes available, the estimation accuracy of minor trip purposes will improve. For such advanced data, the Naïve Bayes classifier would not be suitable because it cannot describe correlations between variables. Hence, more advanced estimation methodologies such as a decision tree, Bayesian network, and support vector machine might be used.

## REFERENCES

Agard, B., Morency, C. and Trépanier, M. 2006. Mining public transport user behaviour from smart card data. *12th IFAC Symp. on Inf. Control Probl. Manuf. – INCOM* 2006, Saint-Etienne in France.

Asakura, Y. and Hato, E. 2004. Tracking survey for individual travel behaviour using mobile communication instruments. *Transp. Res. Part C Emerg. Technol.* 12(3–4): pp. 273-291.

Asakura, Y., Iryo, T., Nakajima, Y., Sugita, K. and Kitano, S. 2008a. TDM experiment of railway and a shopping centre using smart card system. *Proc. TDM Symp.* 2008, Wien in Austria.

Asakura, Y., Iryo, T., Nakajima, Y., Kusakabe, T., Takagi, Y. and Kashiwadani, M. 2008b. Behavioural analysis of railway passengers using smart card data. *Proc. Urban. Transp.* 2008, in Malta.

Asakura, Y., Iryo, T., Nakajima, Y. and Kusakabe, T. 2012. Estimation of behavioural change of railway passengers using smart card data. *Public Transp.* 4(1): pp. 1-16.

Bagchi, M. and White, P.R. 2005. The potential of public transport smart card data. *Transp. Policy* 12(5): pp. 464-472.

Cottrill, C.D., Pereira, F.C., Zhao, F., Dias, I.F., Lim, H.B., Ben-Akiva, M.E. and Zegras, P.C. 2013. Future mobility survey. *Transp. Res. Rec.* 2354: pp. 59 67.

Draijer, G., Kalfs, N. and Perdok, J. 2000. Global positioning system as a data collection method for travel research. *Transp. Res. Rec.* 1719: pp. 147-153.

El Faouzi, N.-E., Leung, H. and Kurian, A. 2011. Data fusion in intelligent transportation systems: progress and challenges – *A Survey. Inf. Fusion* 12(1): pp. 4-10.

Gong, L., Morikawa, T., Yamamoto, T. and Sato, H. 2014. Deriving personal trip data from GPS data: a literature review on the existing methodologies. *Procedia Soc. Behav. Sci.* 138: pp. 557-565.

Hall, D.L. 1992. Mathematical techniques in multisensor data fusion. Artech House, Norwood, MA, USA.

Kamakura, W.A. and Wedel, M. 1997. Statistical data fusion for cross-tabulation. *J. Mark. Res.* 35(4): pp. 485-498.

Kusakabe, T., Iryo, T. and Asakura, Y. 2010. Estimation method for railway passengers' train choice behaviour with smart card transaction data. *Transp.* 37 (5): pp. 731-749.

Kusakabe, T. and Asakura, Y. 2011. Behavioural data mining for railway travellers with smart card data. *Second Int. Workshop on Traffic Data Collect.and Its Stand.*, Brisbane in Australia.

Kusakabe, T. and Asakura, Y. 2014. Behavioural data mining of transit smart card data: A data fusion approach. *Transp. Res. Part C Emerg. Technol.* 46: pp. 179-191.

Kusakabe, T., Seo, T., Goto, H. and Asakura, Y. 2015. Interactive Online Machine Learning Approach for Activity—Travel Survey. *14th int. Conf. Travel Behav. Res.*, Windsor in United Kingdom.

Mitchell, H.B. 2007. *Multi-sensor Data Fusion – An Introduction.* Springer-Verlag, Berlin, Germany.

Morency, C., Trépanier, M. and Agard, B. 2007. Measuring transit use variability with smart-card data. *Transp. Policy* 14(3): pp. 193-203.

Murakami, E. and Wagner, D.P. 1999. Can using global positioning system (GPS) improve trip reporting? *Transp. Res. Part C Emerg. Technol.* 7 (2–3): pp. 149-165.

Pelletier, M., Trépanier, M. and Morency, C. 2011. Smart card data use in public transit: a literature review. *Transp. Res. Part C Emerg. Technol.* 19(4): pp. 557-568.

Rish, I. 2001. An empirical study of the Naïve Bayes classifier. *IJCAI 2001 Workshop on Empir. Methods in Artif. Intell.*, Seattle in USA.

Shen, L. and Stopher, P.R. 2013. A process for trip purpose imputation from Global Positioning System data. *Transp. Res. Part C Emerg. Technol.* 36: pp. 261-267.

Utsunomiya, M., Attanucci, J. and Wilson, N. 2006. Potential uses of transit smart card registration and transaction data to improve transit planning. *Transp. Res. Rec.* 1971: pp. 119-126.

Wolf, J., Guensler, R. and Bachman, W. 2001. Elimination of the travel diary: an experiment to derive trip purpose from GPS travel data. *Transp. Res. Rec.* 1768: pp. 124-134.

## AUTHOR BIOGRAPHY

**Takahiko Kusakabe** is an assistant professor in the University of Tokyo, Japan. He received his Dr. Eng from Kobe University. The focus of his research is behavioural analysis and data mining using passive data set including smart card data of public transport. He uses the data collected by private railway company in the Osaka metropolitan area.

**Yasuo Asakura** is a professor in Tokyo Institute of Technology, Japan. He holds a Dr. Eng from Kyoto University. His study interests of smart card data are performance evaluation of transport system, TDM schemes and ITS application. He uses the data collected by private railway company in the Osaka metropolitan area.

Chapter 6

# A Method for Conducting Before-After Analyses of Transit Use by Linking Smart Card Data and Survey Responses

**C. Brakewood[1,*] and K. Watkins[2]**

## ABSTRACT

Transit agencies are under constant pressure to increase ridership. Many system changes and new technologies, such as making real-time information available, have the potential to increase ridership. However, measuring traveller response to system changes is notoriously difficult. The objective of this chapter is to develop a new method to quantify changes in the number of transit trips due to system changes, and the method is applied to real-time information as an example. The method combines smart card data with survey responses to study the behavior of individual riders before and after the availability of real-time information. First, three conditions are imposed on the joint survey/smart card dataset to assess if each record accurately reflects an individual's travel behavior. The first condition necessitates that the respondent began using real-time information in the appropriate timeframe and had the smart card sufficiently long for the before-after analysis. The second condition tests if one smart card accurately represents one traveller, and the third condition verifies that the smart card record corresponds to the respondent's stated travel behavior. Then, difference of means tests and regression analysis are used to assess changes in monthly transit trips between real-time information users and non-users. In this case, the results suggest that real-time information did not have a significant effect on the number of transit trips in the study; however, the final sample size was small. The primary contribution of this research is the method, which could be repeated to evaluate other transit system changes or technologies.

[1] Civil Engineering Department, City College of New York, 160 Convent Avenue, New York, NY 10031, USA. Email: cbrakewood@ccny.cuny.edu

[2] School of Civil and Environmental Engineering, Georgia Institute of Technology, 790 Atlantic Drive, Atlanta, GA, 30332, USA. Email: kari.watkins@ce.gatech.edu

* Corresponding author

## 1. INTRODUCTION

Because public transportation ridership is affected by numerous factors, many previous studies have had difficulty isolating changes in transit trip-making caused by a particular transit system change, such as the availability of new information and communication technologies. Therefore, the objective of this chapter is to develop a new method using smart card data to quantify differences in transit trips over time due to system changes. While smart card systems are installed for the purpose of revenue collection, the data created from these systems provide a valuable source of information about transit travel over time. Smart card data can be used to assess transit trips before and after a system change; however, in many cases, additional information is needed from the rider beyond the data available in the smart card system. The specific case used in this research to demonstrate the method is the launch of a mobile application that provides real-time information to transit riders. In order to understand which smart card users are also real-time information (RTI) users, the smart card data are combined with the responses from a web-based survey asking about use of RTI. To link the two datasets, a survey question requested the unique smart card number of each respondent. This method was applied to the case study of Atlanta, Georgia.

This chapter proceeds as follows. First, prior research about the use of smart card data to study transit travel behavior is reviewed. The next section provides background information on Atlanta, which is where the data collection and analysis were conducted. Then, the methodology is described and three conditions are applied to the combined smart card/survey dataset. This is followed by the application of the method to evaluate the impacts of RTI on transit travel and last, areas for future research and conclusions are presented.

## 2. LITERATURE REVIEW

This section provides a brief literature review on the uses of smart card data to study travel behavior. Although smart card systems are installed for the purpose of revenue collection, they also provide a rich source of data about transit use (Bagchi and White 2005; Pelletier et al. 2011). Passengers with contactless smart cards pay fares by "tapping" their cards on fareboxes or faregates. With each tap, a record is created that includes the date and time, the type of transaction (boarding, transfer, etc.), fare type, route/station ID, route/line direction, a unique card ID number and possibly other things (Pelletier et al. 2011). Some transit agencies also allow smart card users to register their cards, typically for the purpose of refunding the value of lost/stolen cards or for using autoload features; registration can include a limited amount of personal information.

A growing body of research utilizes transit smart card data and Pelletier et al. (2011) provide a literature review of the uses, which they divide into three groups: operational, tactical and strategic-level applications. Operational-level studies use smart card data to measure various transit supply-and-demand and performance indicators; tactical-level studies commonly focus on service adjustments; and strategic-level studies typically relate to long-term network planning, demand forecasting and travel behavior. In this case, a strategic-level analysis of travel behavior was conducted.

Smart card data have some noteworthy advantages for studying transit travel behavior. Transit providers have found it difficult to examine travel behavior over long timeframes due to a lack of suitable temporal data (Bagchi and White 2005). In contrast, smart card records can be stored for years and accessed as needed over time. Numerous prior studies have taken advantage of the longitudinal nature of smart card records to study the travel patterns of transit riders at the individual level. An early example of this is a study by Morency et al. (2007), who analysed 277 consecutive days of travel on a Canadian transit network and used data mining techniques to study the temporal variability of individuals' transit travel patterns. Another relevant example is an impact analysis of the East London Line, which was a major public transport expansion project that opened in 2010 (Ng 2011). This study used Transport for London's smart card data to create a panel of riders and study the changes in their transit travel behavior patterns due to the opening of this transit line. Similar to this prior study, this analysis will also conduct a before-after study of transit travel behavior; however, it will focus on transit information rather than infrastructure changes.

Another advantage of smart card data is that it is automatically collected and may not be subject to some biases commonly found in self-reported travel data. Unlike travel surveys, the smart card record does not require the traveller to recall or record information about his or her trip patterns, and therefore, could lessen errors in recalling travel behavior (Chapleau et al. 2008).

A major disadvantage of using smart card data to study behavior is a lack of socioeconomic attributes about the cardholder (Pelletier et al. 2011). While some smart card systems collect a limited amount of registration information (Utsunomiya et al. 2006), most lack basic demographic information about the cardholders and none include highly specific attributes, such as the usage of information technologies like mobile applications.

One method to obtain additional data about cardholders is to link smart card data with survey responses by asking for the respondent's smart card number in the survey questionnaire. This was recently done with London's household travel diary (Riegel and Attanucci 2014), a transit origin-destination survey conducted in Santiago, Chile (Munizaga et al. 2014), and a stated preference survey evaluating transit loyalty programs in Shizuoka, Japan (Nakamura et al. 2016). For this research, the procedure of asking for the smart card number was used in a survey that also asked questions about utilization of real-time information, and this joint smart card/survey dataset was then used to evaluate transit travel over time without relying on self-report data to measure trips.

## 3. BACKGROUND

Atlanta was selected as the location for this analysis to demonstrate the proposed methodology. The transit agency in Atlanta is known as the Metropolitan Atlanta Rapid Transit Authority (MARTA), and the smart card system is known as Breeze. Fare media include a plastic contactless Breeze Card, which is most commonly used, and a contactless paper Breeze Ticket, which is primarily used for student tickets, group tickets and special events. A single ride can also be paid directly with cash on buses (MARTA 2014). According to a recent system-wide survey of MARTA riders, over 99% of riders have one or more plastic Breeze Cards (MARTA 2013).

The Breeze system requires tap-in on buses and both tap-in and tap-out on MARTA rail, but this study includes tap-in data only. MARTA riders have the option of registering their Breeze Cards for balance protection and reloading value online; personal information from these processes was not available for this analysis.

Notably, the Breeze Card system was launched in 2006 (Hong 2006), which means that there are nearly ten years of data available to study transit system changes. The system change under consideration in this study is real-time information (RTI) through mobile applications ("apps"), which first became available in late 2013. Therefore, the intervention under study occurred significantly after the implementation of the smart card system, which is necessary for the before-after analysis to evaluate the impacts of RTI on travel behavior using this data source.

## 4. DATA COLLECTION

To assess which MARTA riders use RTI apps, which was the invention under evaluation, a short survey was conducted. The data were collected via a web-based survey, which allowed for questions with images (e.g. a Breeze Card with the smart card number circled and screenshots of RTI apps). The reason for using a web-based survey (as opposed to paper or telephone surveys) was because RTI was primarily accessible via web-enabled devices; therefore, in order to maximize the number of potential respondents who had used RTI, the survey was conducted online.

Responses were collected for one week in early May 2014. Participants were primarily recruited through online channels, including MARTA's social media, the Atlanta Regional Commission email list and other similar email lists. Flyers were also distributed in a small number of train stations to advertise the survey. An incentive of a $5 Starbucks gift card was provided for completing the survey.

### 4.1 Survey Content

The survey was titled "Georgia Tech's Survey of Technologies Used by MARTA Riders" to recruit both users and non-users of RTI, and the survey instrument was divided into five sections. The first section included questions about paying for transit, such as use of a Breeze Card and the respondent's Breeze Card number. The second section contained travel behavior questions, and the third part included questions about mobile RTI. The fourth section asked a few questions about recent MARTA service changes. The last section was composed of socioeconomic questions, including changes during the previous year. Detailed personal information (such as email or home address) was not collected to protect the anonymity of participants at MARTA's request. Last, the survey instrument was reviewed by a dozen Georgia Tech students/staff and a MARTA customer research employee before dissemination.

### 4.2 Responses

A total of 669 participants entered the online survey and of these, 651 respondents answered the first question, which asked how they typically pay for MARTA. Of the 651 respondents, 11 respondents (1.7%) said that they use a paper Breeze Ticket, 7 (1.1%) stated that they pay using cash and 1 respondent (0.2%) was not sure of the fare media that s/he typically uses. This left 632 survey respondents who use one or more Breeze Cards and of those, 538 provided a smart card number. The smart card numbers were provided to MARTA, and 497 matched active Breeze Card numbers. Three additional participants were removed due to restrictions (i.e. under age 18), so the remaining sample size was 494, or 73.8% of all those who entered the survey. Then, the smart card records for the 494 eligible participants were merged with the corresponding survey responses using the smart card number.

Because this survey was collected with non-probability sampling methods, questions about socioeconomic status and basic travel behavior were asked to understand the representativeness of the sample. Table 1 shows summary statistics of these survey questions for the 494 eligible participants and then compares them to MARTA's most recent system-wide survey (MARTA 2013). As can be seen in the table, the respondents to this survey differed from typical MARTA riders in a few noteworthy ways; this study includes more participants who were Caucasian, had higher income levels and took fewer transit trips per week than typical MARTA riders.

Last, it should be noted that smart card records were aggregated to the number of trips per day per mode (bus/rail) by MARTA and the complete trip history (i.e. time-stamped tap-in locations) was not provided as a safeguard to protect the privacy of respondents at MARTA's request. Also, to ensure that the records from the Breeze Card database were accurate, the smart card records of a few Georgia Tech researchers were requested and reviewed.

**Table 1.** Characteristics of study participants compared with system-wide MARTA riders

| | | Study Participants | | MARTA Riders* |
|---|---|---|---|---|
| **Category** | **Variable** | **Count** | **% Column** | **% Column** |
| Total | All Respondents | 494 | 100% | 100% |
| Gender | Male | 246 | 49.8% | 50.7% |
| | Female | 232 | 47.0% | 49.3% |
| | No Answer | 16 | 3.2% | - |
| Age | Age 18-24 (0-24 for all MARTA riders**) | 62 | 12.6% | 23.3% |
| | Age 25-34 | 229 | 46.4% | 25.9% |
| | Age 35-44 | 113 | 22.9% | 17.5% |
| | Age 45-54 | 56 | 11.3% | 18.4% |
| | Age 55-64 | 19 | 3.8% | 11.8% |
| | Age 65 and older | 3 | 0.6% | 3.1% |
| | No Answer | 12 | 2.4% | - |
| Annual Household Income | Under $10,000 | 20 | 4.0% | 19.9% |
| | $10,000 to $19,999 | 28 | 5.7% | 19.2% |
| | $20,000 to $29,999 | 48 | 9.7% | 20.5% |
| | $30,000 to $39,999 | 34 | 6.9% | 12.6% |
| | $40,000 to $49,999 | 40 | 8.1% | 7.2% |
| | $50,000 to $74,999 | 83 | 16.8% | 9.1% |
| | Over $75,000 | 212 | 42.9% | 11.4% |
| | No Answer | 29 | 5.9% | - |
| Spanish, Hispanic or Latino Descent | Yes, Hispanic | 20 | 4.0% | 6.4% |
| | No, not Hispanic | 461 | 93.3% | 93.6% |
| | No Answer | 13 | 2.6% | - |
| Ethnicity | American Indian or Alaska Native | 2 | 0.4% | 0.2% |
| | Asian (includes Asian Indian) | 40 | 8.1% | 3.0% |
| | Black or African American | 57 | 11.5% | 76.3% |
| | White | 368 | 74.5% | 15.4% |
| | Other | 12 | 2.4% | 5.1% |
| | No Answer | 15 | 3.0% | - |
| Number of One-way MARTA Trips in the Last Week (Bus and Train) | 0 to 4 trips | 291 | 58.9% | 34.2% |
| | 5 to 8 trips | 48 | 9.7% | 17.9% |
| | 9 or more trips | 153 | 31.0% | 47.9% |
| | No Answer | 2 | 0.4% | - |

* System-wide statistics from MARTA's FY13 Quality of Service Survey Annual Report.

** Questions not equivalent. MARTA's system-wide survey included respondents age 0-18, but this study did not.

## 5. METHODOLOGY

This section describes the method used to evaluate the combined smart card/survey dataset. First, the use of the RTI, which was the intervention under evaluation, was considered. Next, three conditions were investigated to assess if each record in the smart card/survey dataset accurately reflects an individual's travel behavior. The first condition necessitates that the person began using RTI in the appropriate timeframe and had the smart card sufficiently long for the before-after analysis. The second condition tests if one smart card represents one traveller. The third condition verifies that the smart card record corresponds to the respondent's stated travel behavior.

### 5.1 The Intervention: Availability of Real-Time Information

To assess the intervention, which was the availability of RTI, the survey contained questions in which the respondent was presented with images of the most popular RTI apps in Atlanta and was asked if s/he has used RTI. A total of 302 of the 494 eligible participants (61%) used one or more RTI apps, and they compose the user group. Respondents who stated that they had not used one or more RTI apps were categorized as non-users.

### 5.2 Condition 1: Panel Eligibility

The first condition imposed on the joint smart card/survey dataset was that of *panel eligibility*. For the before-after analysis, a month before the main release of RTI in Atlanta (April 2013) and the same month one year later (April 2014) were selected because the intervention (the launch of various RTI apps) occurred at different times in 2013 and 2014. Since there was a small possibility that respondents began using RTI during the "before" period (April 2013 or earlier) or in the middle of the "after" period (April 2014), respondents were asked to recall when they began using RTI. Similarly, to ensure that the smart card was active for the entire study period, respondents were asked to recall when they acquired their smart card.

#### 5.2.1 *Condition 1A: Panel Eligibility of the Intervention*

First, respondents were asked to recall when they started using an app with RTI (i.e. the intervention), and the results are shown in Table 2. Most respondents (201 RTI users) began using the apps between May 2013 and March 2014 and were deemed panel eligible. Another 36 could not recall when they began using RTI and 2 did not answer the question, and it was assumed that they began within the last year. Therefore, a total of 239 respondents were deemed panel eligible RTI users, and they could be compared to the 192 non-users. This resulted in a sample size of 431 respondents meeting Condition 1A.

**Table 2.** Condition 1A (Panel eligibility of the intervention)

| Survey Question: When did you start using an app with RTI? | Meet 1A | Total | Percent |
|---|---|---|---|
| Began using RTI before April 2013 | No | 37 | 7% |
| Began between May 2013 and March 2014 | Yes | 201 | 41% |
| April 2014 or later | No | 26 | 5% |
| Cannot remember | Yes | 36 | 7% |
| No Answer | Yes | 2 | 0% |
| RTI User Total | 239 | 302 | 61% |
| Non-users | Yes | 192 | 39% |
| Grand Total | 431 | 494 | 100% |

### 5.2.2 *Condition 1B: Panel Eligibility of the Smart Card*

Respondents were asked if they got their Breeze Card within the last year or more than a year ago, and the results are shown Table 3. Of the 431 respondents meeting Condition 1A, a total of 264 respondents (61%) stated that they have had their Breeze Card for more than a year. Another 41 respondents (10%) could not recall when they acquired their Breeze Card, and it was assumed that these respondents were also panel eligible. This resulted in a total of 305 participants who met Condition 1B. These survey responses were also compared to the smart card record for April 2013, which is shown in Table 3. Notably, this condition excludes any person(s) who began riding transit within the last year, since they did not have a Breeze Card a year ago.

**Table 3.** Condition 1B (Panel eligibility of the smart card)

| Survey Question: When did you get your Breeze Card? | Meet 1B | Breeze Card Data | | | |
|---|---|---|---|---|---|
| | | No Trips in April 2013 | 1+ Trips in April 2013 | Total | Percent |
| Within the last year | No | 111 | 15 | 126 | 29% |
| One year or more ago | Yes | 111 | 153 | 264 | 61% |
| I'm not sure | Yes | 29 | 12 | 41 | 10% |
| Total | 305 | 251 | 180 | 431 | 100% |

## 5.3 Condition 2: Completeness and Uniqueness (One Smart Card = One Person)

Next, each record was tested for *completeness* and *uniqueness*. A Breeze Card record was considered *complete* if the respondent did not use any other form of payment when riding MARTA; consequently, all of the respondent's transit trips would be captured in the smart card record. A Breeze Card was considered *unique* if it was only used by a single person. A Breeze Card record would not be unique if it is shared with others because this represents the travel behavior of more than one person. If both the conditions of completeness and uniqueness are met, it was assumed that one smart card represents one person. These conditions were assessed using the responses to three survey questions.

### 5.3.1 Condition 2A: Complete with One Breeze Card

The first survey question pertaining to completeness asked if a respondent had one Breeze Card or two or more Breeze Cards. As is shown in Table 4, 86 (71+15) respondents who met Condition 1B have two or more Breeze Cards, and therefore, their smart card records may not be complete. The remaining 219 (193+26) participants were assumed to meet Condition 2A.

### 5.3.2 Condition 2B: Complete with No Other Fare Media

As a second measure of completeness, all participants were asked if they pay for MARTA in other ways (such as cash or a paper ticket). Table 4 shows that a total of 193 participants who met Condition 2A use only their Breeze Card and were deemed complete.

### 5.3.3 Condition 2C: Unique

Finally, to understand uniqueness, survey respondents were asked if they share their Breeze Card, and to what extent they share their card (e.g. occasionally, often). A total of 159 respondents who met condition 2B also met the uniqueness condition because they stated that they never share their single Breeze Card (Table 4). Therefore, it was assumed that the smart cards of those 159 respondents accurately represent the transit travel of only those respondents.

**Table 4.** Conditions 2A, 2B and 2C (Completeness and uniqueness)

| Survey Question: Unique | Survey Questions: Complete | | | | |
|---|---|---|---|---|---|
| | 1 Breeze Card | | 2+ Breeze Cards | | Total |
| | Uses only Breeze Card | Uses Cash or Ticket | Uses only Breeze Card | Uses Cash or Ticket | |
| I never share my Breeze Card (1 or 2 cards) | 159 | 20 | 42 | 8 | 229 |
| I have shared my Breeze Card once or twice | 25 | 4 | 14 | 3 | 46 |
| I occasionally share my Breeze Card | 3 | 2 | 13 | 4 | 22 |
| I often share my Breeze Card | 4 | 0 | 1 | 0 | 5 |
| I'm not sure | 1 | 0 | 0 | 0 | 1 |
| Other | 1 | 0 | 1 | 0 | 2 |
| Total | 193 | 26 | 71 | 15 | 305 |

## 5.4 Condition 3: Congruence (That Smart Card = That Person)

Last, the condition of *congruence* was assessed by comparing each smart card record to a self-reported travel behavior survey question to ensure that the smart card record represents that particular person. The purpose of this was to identify errors when the respondent entered his or her smart card number in the survey or potential errors in the Breeze Card system.

The specific method to assess congruence was comparing the number of MARTA train trips made in the last week from the smart card record to a self-reported survey question. Each survey respondent was instructed to begin counting train trips from the previous day and continuing back seven days. Because each online survey response included a time and date of completion, the self-reported number of MARTA train trips was compared to the same seven days of the smart card record. Respondents were also instructed to count train-to-train transfers as single trips, but transfers that involved bus modes (bus and train) were counted separately. This was to ensure that the number of "taps" in the smart card database aligned with self-reported trips, since bus-to-train transfers involving tapping the smart card at the transfer point whereas train-to-train transfers do not (since one stays within the fare gates).

### 5.4.1 Condition 3A: Closely Congruent

As is shown in Table 5, 135 respondents (of those who were met Condition 2C) had self-reported trips that matched the respective smart card trip history within two train trips. These survey responses were deemed to be "closely congruent" with the respective smart card. "Close" congruence was considered because self-reported travel behavior questions are often subject to error, particularly recall bias in which respondents cannot correctly remember their travel (Stopher 2012). There is also the possibility that a transaction was missing from the smart card dataset, since prior research has identified this as a possible flaw with smart card data (Utsunomiya et al. 2006).

### 5.4.2 Condition 3B: Perfectly Congruent

Table 5 shows that 100 respondents (of those who met all other conditions) had survey responses that perfectly matched the respective smart card record and were deemed "perfectly congruent."

## 5.5 Summary

After imposing conditions on the joint survey/smart card dataset, 100 (20%) of the 494 eligible participants were found to meet all three conditions. Table 6 shows the sample size as each condition was applied. Since the

**Table 5.** Conditions 3A and 3B (Closely and perfectly congruent)

| Survey Question: Number of Train Trips in the Last 7 Days | Breeze Card Data: Number of Responses | | | | |
|---|---|---|---|---|---|
| | Closely Congruent | | Perfectly Congruent | | |
| | Count | % Total | Count | % Total | Total |
| 0 trips | 63 | 100% | 62 | 98% | 63 |
| 1 trips | 11 | 100% | 7 | 64% | 11 |
| 2 trips | 17 | 94% | 8 | 44% | 18 |
| 3 trips | 0 | – | 0 | – | 0 |
| 4 trips | 10 | 77% | 5 | 38% | 13 |
| 5 trips | 2 | 50% | 0 | 0% | 4 |
| 6 trips | 0 | 0% | 0 | 0% | 1 |
| 7 trips | 0 | 0% | 0 | 0% | 2 |
| 8 trips | 4 | 57% | 3 | 43% | 7 |
| 9 trips | 0 | – | 0 | – | 0 |
| 10 trips | 16 | 76% | 7 | 33% | 21 |
| 11 trips or more | 12 | 63% | 8 | 42% | 19 |
| Total | 135 | – | 100 | – | 159 |
| Percent Total | 85% | – | 63% | – | 100% |

**Table 6.** Conditions and sample sizes

| Number | Condition | Sample Size | Percent Total |
|---|---|---|---|
| – | Full survey/smart card data set | 494 | 100% |
| 1A | Panel eligibility of the intervention | 431 | 87% |
| 1B | Panel eligibility of the smart card | 305 | 62% |
| 2A | Complete with one breeze card | 219 | 44% |
| 2B | Complete with no other fare media | 193 | 39% |
| 2C | Unique | 159 | 32% |
| 3A | Closely congruent | 135 | 27% |
| 3B | Perfectly congruent | 100 | 20% |

sample size decreased substantially, all conditions were considered in the following analysis.

## 6. EVALUATION OF THE INTERVENTION

Next, the joint smart card/survey dataset was used in a before-after analysis of the intervention, which was the availability of RTI in this example. This analysis is divided into two parts. The first section presents simple statistics to compare the number of transit trips by RTI users with non-users, and the second section uses regression analysis to control for other factors that may be influencing transit travel.

## 6.1 Difference of Mean Differences

The first analysis compares the number of transit trips from the smart card data before and after the availability of RTI for users and non-users. The period of analysis was four weeks in April 2013 and April 2014 beginning with the first Tuesday of the month so that there were the same number and type of days in each period (i.e. 4 Mondays, 4 Tuesdays, etc.). Conveniently, April also includes typical school trips (the local universities are all in session) and no major holidays.

Table 7 shows the mean (M), median (Med), standard deviation (SD), minimum (Min), and maximum (Max) number of transit trips for the four weeks in April 2013 and April 2014 broken down by RTI users versus non-users. The difference between April 2013 and 2014 was calculated for each individual, and this difference was used in a difference of means test between RTI users and non-users. The results are shown for the entire dataset (n=494) in the leftmost column of Table 7. Each condition (1A to 3B) was progressively applied moving toward the right of the table, and a comparable analysis was conducted.

When the full dataset (n=494) is considered, the results suggest that RTI users increased transit trips significantly more than non-users from April 2013 to April 2014 ($mean\ difference_{RTI\text{-}users}=11.7$ *trips*, $mean\ difference_{non\text{-}users}=4.9$ *trips, two-tailed* $p\text{-}value=0.0006$). There are similar findings when Condition 1A (Panel Eligibility of the Intervention) is applied. When Conditions 1B to 3B are applied, the mean difference in trips from April 2013 to 2014 for the RTI user group is still a greater increase than that of the non-user group; however, this difference is not statistically significant. This could be because the more filtered datasets have smaller sample sizes and therefore have larger variances of the estimator, making it more difficult to detect a difference. It may also be because RTI users took, on average, more trips in April 2013 than non-users, which suggests that those who use transit more were more likely to adopt RTI. Difference of means tests were also run for each mode (bus, rail) separately, and similar results were found in which RTI was only significant for the full dataset and Condition 1A.

Two important notes should be made about this analysis. First, when examining the median difference in trips (as opposed to the mean), there were limited changes from April 2013 to April 2014 regardless of which conditions were applied. Second, system-wide MARTA ridership over the study period was relatively stable; there were 129.9 million unlinked MARTA trips in fiscal year 2013 and 129.1 million in fiscal year 2014 (MARTA 2015). Also, system-wide ridership figures for only the months of interest (April 2013 and April 2014) reveal a slight decrease in ridership over the study period (11.5 million unlinked passenger trips in April 2013 versus

**Table 7.** Before-After analysis of transit trips

| | | All Data (Matches) | | Condition 1A (Panel Eligible) | | Condition 1B (Panel Eligible) | | Condition 2A (Complete) | | Condition 2B (Complete) | | Condition 2C (Unique) | | Condition 3A (Congruent) | | Condition 3B (Congruent) | |
|---|---|---|---|---|---|---|---|---|---|---|---|---|---|---|---|---|---|
| RTI Use | | User | No | User | No | User | No | User | No | User | No | User | No | User | No | User | No |
| Count | | 302 | 192 | 239 | 192 | 166 | 139 | 114 | 105 | 99 | 94 | 77 | 82 | 60 | 75 | 38 | 62 |
| April 2013 | M | 10.2 | 4.7 | 10 | 4.7 | 12.9 | 6.2 | 14.1 | 6.8 | 15.8 | 7.4 | 17.5 | 8.4 | 15.6 | 5.7 | 12.8 | 4.1 |
| | Med | 0 | 0 | 0 | 0 | 2 | 0 | 2 | 0 | 3 | 0 | 5 | 1 | 3 | 0 | 0.5 | 0 |
| | SD | 20.2 | 14.5 | 19.1 | 14.5 | 20.1 | 16.5 | 20.3 | 18 | 21.2 | 18.9 | 22 | 20 | 21.7 | 12.3 | 22.2 | 9.4 |
| | Min | 0 | 0 | 0 | 0 | 0 | 0 | 0 | 0 | 0 | 0 | 0 | 0 | 0 | 0 | 0 | 0 |
| | Max | 113 | 138 | 113 | 138 | 91 | 138 | 91 | 138 | 91 | 138 | 91 | 138 | 91 | 59 | 91 | 46 |
| April 2014 | M | 21.9 | 9.6 | 21.4 | 9.6 | 21.2 | 10.1 | 21.4 | 11.9 | 21.7 | 12.2 | 22.8 | 12.5 | 21.7 | 7.9 | 21.1 | 5.1 |
| | Med | 8.5 | 1 | 6 | 1 | 5 | 1 | 6 | 1 | 9 | 1 | 12 | 1 | 7.5 | 1 | 3 | 0 |
| | SD | 29.3 | 22.4 | 29.7 | 22.4 | 31.1 | 23.8 | 27.4 | 26.6 | 26.9 | 26.5 | 27.6 | 27 | 27.5 | 14.7 | 29.8 | 10.6 |
| | Min | 0 | 0 | 0 | 0 | 0 | 0 | 0 | 0 | 0 | 0 | 0 | 0 | 0 | 0 | 0 | 0 |
| | Max | 212 | 205 | 212 | 205 | 212 | 205 | 112 | 205 | 112 | 205 | 112 | 205 | 112 | 70 | 112 | 40 |
| Difference | M | 11.7 | 4.9 | 11.4 | 4.9 | 8.3 | 3.9 | 7.3 | 5.1 | 5.9 | 4.8 | 5.2 | 4 | 6.1 | 2.2 | 8.3 | 1 |
| | Med | 2 | 0 | 1 | 0 | 0 | 0 | 0 | 0 | 0 | 0 | 0 | 0 | 0 | 0 | 0.5 | 0 |
| | SD | 27.8 | 15.8 | 28.3 | 15.8 | 29.1 | 15.7 | 24.6 | 17.9 | 23.2 | 16.3 | 24.3 | 14.7 | 25.4 | 11.3 | 25.1 | 8.9 |
| | Min | -51 | -32 | -51 | -32 | -51 | -32 | -44 | -32 | -44 | -32 | -44 | -32 | -24 | -32 | -17 | -32 |
| | Max | 174 | 95 | 174 | 95 | 174 | 95 | 112 | 95 | 112 | 80 | 112 | 67 | 112 | 45 | 112 | 40 |
| | | t = -3.478 | | t = -3.016 | | t = -1.69 | | t = -0.7524 | | t = -0.369 | | t = -0.3728 | | t = -1.097 | | t = -1.732 | |
| | | p=0.0006 | | p=0.003 | | p=0.092 | | p=0.453 | | p=0.713 | | p=0.710 | | p=0.276 | | p=0.0905 | |

10.9 million trips in April 2014). Therefore, the limited change in median number of trips from April 2013 to 2014 in this study appears consistent with system-wide MARTA ridership trends.

## 6.2 Regression Analysis

Since changes in an individual's monthly transit trips could be attributed to factors other than the intervention (RTI), survey respondents were asked a series of retrospective questions to understand possible changes that may have influenced their travel behavior between April 2013 and 2014. This included questions about changes in household size, automobile ownership, job location and household location over the last year. Additionally, a few questions about awareness of MARTA's minor service changes that occurred in December 2013 were included in the survey instrument, since this could have also caused changes in transit travel during the study period. The results of these questions were then included in a regression model to assess the impact of the intervention while controlling for these other factors. The dependent variable in the regression was the difference in monthly trips (precisely, four weeks) from 2013 to 2014 from the smart card record, and the independent variables included the previously mentioned retrospective survey questions, as well as standard socioeconomic characteristics (e.g. ethnicity, age, etc.).

Various regression models were assessed, and the models selected for presentation are shown in Table 8. These models contain only variables that were consistently significant across datasets or significant in the dataset that met the congruence conditions (3A and 3B). The variable of interest, RTI, was only significant in the regression models using the full dataset and the dataset in which Condition 1A was met. When the additional conditions were imposed, RTI use was no longer significant. The other variables that were consistently significant as the sample size decreased were having a valid driver's license, which was associated with a decrease in MARTA trips from 2013 to 2014, and being African American, which was associated with an increase in transit trips. However, both of these variables were to some extent collinear with the intercept: only 9% of the final sample was African American and 96% had a driver's license. Two other variables were significant in some of the models. Increasing the number of cars in a household during the previous year was associated with a decrease in MARTA trips in the full dataset and when condition 1A (Panel Eligibility of the Intervention) was applied. On the other hand, awareness of MARTA's recent (minor) service change was associated with an increase in trips in the models when the congruence conditions (3A and 3B) were applied. This suggests that the minor service changes in December 2013 positively impacted the number of trips riders made on MARTA, although further study of this is recommended. Last, the goodness of fit was limited for all models: there was an R-squared of 0.15

**Table 8.** Regression analysis of difference in transit trips

| | All Data | Condition 1A | Condition 1B | Condition 2A | Condition 2B | Condition 2C | Condition 3A | Condition 3B |
|---|---|---|---|---|---|---|---|---|
| | (Matches) | (Panel Eligible) | (Panel Eligible) | (Complete) | (Complete) | (Unique) | (Congruent) | (Congruent) |
| Intercept | 20.887<br>(5.644)*** | 20.786<br>(6.069)*** | 25.676<br>(8.829)*** | 34.222<br>(10.328)*** | 40.491<br>(11.043)*** | 37.468<br>(11.355)*** | 37.115<br>(14.754)** | 36.146<br>(16.956)** |
| Use Real-Time Information | 6.61<br>(1.897)*** | 6.494<br>(2.112)*** | 4.047<br>(2.627) | 1.552<br>(2.461) | 0.331<br>(2.265) | −0.668<br>(2.606) | −0.664<br>(2.526) | 2.651<br>(3.04) |
| Has a Licence | −18.633<br>(5.886)*** | −18.183<br>(6.358)*** | −25.356<br>(9.059)*** | −34.277<br>(10.622)*** | −40.958<br>(11.344)*** | −38.444<br>(11.775)*** | −38.944<br>(15.191)** | −38.436<br>(17.662)** |
| African American | 16.544<br>(5.797)*** | 13.723<br>(6.405)** | 15.499<br>(7.590)** | 19.674<br>(7.859)** | 15.727<br>(7.135)** | 17.589<br>(8.436)** | 18.47<br>(9.266)** | 10.815<br>(9.45) |
| Increased Cars in Household | −8.215<br>(2.488)*** | −8.015<br>(2.582)*** | −6.785<br>(2.862)** | −3.568<br>(2.805) | −1.072<br>(2.841) | −2.159<br>(2.911) | −4.237<br>(2.393)* | −2.159<br>(2.305) |
| Aware of Service Change | 0.012<br>(2.15) | −0.083<br>(2.342) | 1.958<br>(2.776) | 4.22<br>(2.608) | 4.604<br>(2.365)* | 4.565<br>(2.571)* | 6.231<br>(2.819)** | 6.647<br>(3.056)** |
| $R^2$ | 0.15 | 0.13 | 0.16 | 0.26 | 0.32 | 0.33 | 0.35 | 0.30 |
| Observations^ | 477 | 416 | 296 | 214 | 189 | 155 | 131 | 98 |

* p<0.1; **p<0.05; ***p<0.01;

^ Number of observations reduced due to missing responses for specific questions. Values shown in parentheses are robust standard errors.

for the full dataset and this increased to 0.30 when all of the conditions (1A-3B) were applied.

## 7. AREAS FOR IMPROVEMENT AND FUTURE RESEARCH

This section discusses potential improvements and challenges for future research. First, the survey responses were collected via non-probability sampling and were not representative of all MARTA riders. This survey substantially differed from MARTA's last system-wide survey in three ways: there were more participants who were Caucasian, had higher income levels, and took fewer transit trips per week than typical MARTA riders (MARTA 2013). Although regression analysis was performed to help control for these differences, use of this methodology to test future transit system changes should use probability sampling to increase the generalizability of descriptive statistics.

Another possible improvement is incorporating assumptions pertaining to the "shrinking" sample size in the sampling plan. The original dataset began with 494 records, but this decreased to 100 (20%) after strictly imposing the three conditions. Future applications of this method should increase the sampling rate in anticipation of this.

A third future enhancement is including a survey question asking if a person began riding transit in the last year. New riders were not considered in this analysis, since these respondents did not have smart cards in the "before" period (Condition 1B). However, it is possible that new riders began using transit because of the availability of RTI, which was the intervention under evaluation.

Another improvement pertains to the condition of congruence, which compared the number of train trips in the last week from the survey to the smart card records. Survey questions are often subject to error, particularly recall bias in which respondents cannot correctly remember their travel (Stopher 2012). Perhaps a better measure of congruence is "home" station, since respondents are likely able to recall this more easily. Additionally, requesting a respondent's smart card number twice to avoid unintentional errors when entering the number could improve congruence condition tests.

A potential challenge for future research is consistency using smart card "taps" to measure transit trips over time if there are fare policy changes. This study was conducted during a timeframe when there were no known changes in fare policy, but shortly afterward, MARTA changed their bus open door policy at transfer locations, which could impact the number of "taps" in future analyses.

Last, a noteworthy challenge to applying this methodology more broadly is privacy concerns on behalf of the transit agency regarding smart card data (Dempsey 2008). Transit agencies may be hesitant (or completely unwilling) to share their data with researchers, particularly if they have stringent privacy policies. For this research, the transit agency was willing to share data in an aggregated form (smart card taps per day); however, there are numerous additional analyses that could be performed if complete smart card trip histories (including time-stamped tap-ins by station/route) are made available to researchers, such as assessing changes in transit usage by time of day or by frequency of service.

## 8. CONCLUSIONS

In this chapter, a methodology was developed to combine smart card data with survey responses to evaluate changes in transit travel over time. This method was applied to an empirical analysis of real-time information (RTI) in Atlanta. First, three conditions were imposed on the joint smart card/survey dataset, which reduced the sample size to 20% of the original dataset. Then, difference of mean tests and regression analysis were used to compare changes in monthly transit trips from April 2013 to April 2014 between RTI users and non-users. The results for the larger initial dataset suggest that RTI was associated with an increase in transit trips. However, when the conditions were applied and the sample size was reduced, the difference in trips was not significantly different between RTI users and non-users. This may because RTI users took, on average, more trips in April 2013 than non-users, which suggests that those who use transit more were more likely to adopt RTI.

A primary contribution of this research is the method to combine smart card data with survey responses to evaluate changes in transit travel. Traditional surveys lack a method of accurately measuring travel over extended periods of time (unless surveys are repeated) and the smart card dataset advantageously provides a record of transit trips needed for before-after or panel analyses. Similarly, the survey instrument can be used to gather socioeconomic information and other characteristics of the respondent, which would otherwise be unavailable in a smart card dataset. This methodology could be used to evaluate other transit system changes – beyond RTI – and more broadly applied for transit marketing and travel behaviour analyses in the future. Planners and market researchers conducting regular transit customer satisfaction surveys could include a few additional questions about smart cards – particularly the smart card number – and apply this methodology to evaluate the impacts of other system, policy, or planning changes on transit travel.

## ACKNOWLEDGEMENTS

This research was funded by an US DOT Eisenhower fellowship, Georgia Tech's GVU Center and the National Center for Transportation Systems Productivity and Management (NCTSPM) University Transportation Center (#2013-042). The authors would like to acknowledge MARTA for providing the smart card data. Thanks to Gregory Macfarlane for introducing the terminology of congruence and to the Georgia Tech students who helped launch RTI in Atlanta, particularly Landon Reed, Tushar Humbe, Derek Edwards and Aaron Gooze.

## REFERENCES


Bagchi, M. and White, P.R. 2005. The Potential of Public Transport Smart Card Data. *Transport Policy*, 12(5), pp. 464-474.

Chapleau, R., Trépanier, M. and Chu, K.K. 2008. The ultimate survey for transit planning: Complete information with smart card data and GIS. *In 8th International Conference on International Steering Committee for Travel.* Survey Conferences, Lac d'Annecy, France.

Dempsey, S. 2008. Privacy Issues with the Use of Smart Cards. *Legal Research Digest*, 25. Transit Cooperative Research Program of the Transportation Research Board of the National Academies, Washington, D.C.

Hong, Y. 2006. Transition to Smart Card Technology: How Transit Operators Can Encourage the Take-Up of Smart Card Technology. *Master's Thesis, Massachusetts Institute of Technology*, Cambridge, MA.

Metropolitan Atlanta Rapid Transit Authority (MARTA). 2013. FY13 *Quality of Service Survey Annual Report.* Unpublished internal report.

Metropolitan Atlanta Rapid Transit Authority (MARTA). 2014. *Fares and Passes.* Retrieved from http://www.itsmarta.com/fares-passes.aspx. Accessed May 8, 2014.

Metropolitan Atlanta Rapid Transit Authority (MARTA). 2015. FY2015 *Performance Measure Update.* Unpublished internal spreadsheet.

Morency, C., Trepanier, M. and Agard, B. 2007. Measuring Transit Use Variability with Smart-Card Data. *Transport Policy*, 14, pp. 193-203.

Munizaga, M., Devillaine, F., Navarrete, C. and Silva, D. 2014. Validating Travel Behaviour Estimated from Smartcard Data, *Transportation Research Part C: Emerging Technologies*, 44, pp. 70-79.

Nakamura, T., Uno, N., Nakamura, N., Schmöcker, J.-D. and Iwamoto, T. 2016. Urban Public Transport Mileage Cards: Analysis of their potential with smart card data and an SP survey. *Compendium of the 95th Annual Meeting of the Transportation Research Board.* Washington, D.C.

Ng, A. 2011. Use of Automatically Collected Data for the Preliminary Impact Analysis of the East London Line Extension. *MS Thesis*. Massachusetts Institute of Technology, Cambridge, MA.

Pelletier, M.P., Trépanier, M. and Morency, C. 2011. Smart Card Data Use in Public Transit: A Literature Review. *Transportation Research Part C: Emerging Technologies*, 19(4), pp. 557-568.

Riegel, L. and Attanucci, J. 2014. Using Automatically Collect Smart Card Data to Enhance Travel Demand Surveys. *Proceedings of the Transportation Research Board 93rd Annual Meeting*, Washington, D.C.

Stopher, P. 2012. Collecting, Managing and Assessing Data Using Sample Surveys. Cambridge University Press, 149.

Utsunomiya, M., Attanucci, J. and Wilson, N. 2006. Potential Uses of Transit Smart Card Registration and Transaction Data to Improve Transit Planning. *Transportation Research Record: Journal of the Transportation Research Board*, No. 1971, Transportation Research Board of the National Academies, Washington, D.C., pp. 119-126.


## AUTHOR BIOGRAPHY

**Dr. Candace Brakewood** is an Assistant Professor of Civil Engineering at the City College of New York. She has a PhD in Civil Engineering from Georgia Institute of Technology, dual Master of Science degrees in Transportation and Technology Policy from Massachusetts Institute of

Technology, and a Bachelor of Science in Mechanical Engineering from Johns Hopkins University. Her research focuses on understanding how new information and communication technologies can be used to improve public transportation systems. She has extensive experience working with smart card data and mobile ticketing data.

**Dr. Kari Watkins** returned to Georgia Tech, her undergraduate alma mater, to become a faculty member in 2011 after completing her PhD at the University of Washington. Dr. Watkins' dissertation involved co-creating the award-winning OneBusAway program to provide transit information tools and assess their impacts on riders, winning her the Council of University Transportation Centers (CUTC) Wootan Award for best dissertation in transportation policy and planning. Her teaching and research interests include multi-modal transportation planning, the use of technology in transportation, traveller information and complete streets design.

Chapter 7

# Multipurpose Smart Card Data: Case Study of Shizuoka, Japan

**T. Nakamura[1,*], N. Uno[2], J-D. Schmöcker[3] and T. Iwamoto[4]**

**ABSTRACT**

This chapter discusses whether it is possible to use smart cards, not only for better understanding of supply and demand patterns but also for directly influencing travel behaviour. For this, a case study with smart card data obtained from Shizuoka, Japan is conducted. The smart card belongs to Shizutetsu group, a private company operating transport services as well as retail facilities. The group introduced a point loyalty scheme rewarding frequent public transport usage with points that could be redeemed for travel or shopping. With a SP survey it is analysed how far customers could be persuaded to change their behaviour if the point system is changed and potential large amount of points could be obtained via lottery like bonus points. It is found that there is some potential to attract customers to use public transport by such a scheme, though it is mostly existing users that would travel more if the point scheme is attractive. Attracting non-public transport users by loyalty points appears difficult.

## 1. INTRODUCTION

The primary purpose of smart cards is to enable cash-free payments for public transport. This creates advantages for the users as well as operators:

[1] Department of Urban Management, Graduate School of Engineering, Kyoto University, Rm 434, C-1-2, Kyoto-Daigaku Katsura Campus, Nishikyo, Kyoto, 615-8540, JAPAN. Email: nakamura@trans.kuciv.kyoto-u.ac.jp

[2] Department of Urban Management,Graduate School of Engineering, Kyoto University. Email: uno@trans.kuciv.kyoto-u.ac.jp

[3] Department of Urban Management,Graduate School of Engineering, Kyoto University. Email: schmoecker@trans.kuciv.kyoto-u.ac.jp

[4] Shizuoka Railway Co. LTD. Email: t.iwamoto@shizutetsu.co.jp

* Corresponding author

For travellers it is convenient as they do not have to spend time looking up fare tables, buying tickets and dealing with coins. For operators the reduced cash handling, paper ticket printing and station crews are also helpful. More importantly though, especially for bus services, faster boarding and alighting time leads also to improve service regularity and to reduce operational costs. Moreover, there are "indirect" benefits of smart cards as the data obtained through these can possibly replace expensive surveys and replace technologies such as automatic passenger count systems. As discussed in several chapters in this book, the data analysis can then support the service planning at various stages. This chapter suggests that smart cards can also be used more directly to influence travel behaviour if point systems are introduced.

Point systems are known to encourage user loyalty (Michael, 2004). Credit card companies and stores of all kinds provide users with member cards that allow them to pay with it and at the same time to collect points. Literature such as Davis (1999) reports that smart cards have an impact on shopping behaviour. In his UK example, it is reported that cardholders spend 10% more in drug stores. Considering transport applications, airline mileage points have also become a major attraction for many customers influencing their behaviour. The mileage systems do not only influence the choice of airline alliances but also mode choice (e.g., in Japan sometimes customers take a flight instead of the Shinkansen to collect mileage points) and possibly even trip generation.

Mileage point systems can usually be exchanged for extra travel as well as for goods (Taylor and Neslin, 2005). Such point systems are much less applied for urban public transport for a number of reasons. First, often there is no strong competition between operators and second, if there is, generally the low price of public transport tickets does not seem to justify such a point system and third, until recently, tickets were mostly purchased with cash and therefore tracking travel records is cumbersome. Recently, the third point has become less of an issue in many cities due to the introduction of smart card systems. The first and second issues though appear to remain obstacles.

One application of public transport point systems is to introduce them not in order to encourage loyalty but to encourage sustainable behaviour. In Nagoya, Japan, a case study is conducted where 5% of the fares were given back to the users as "transport eco-points". Sato et al. (2006) report that this increased the rail trips temporarily by up to 5.6%. In general Morikawa (2008) suggests that providing users with eco-points for environment friendly behaviour can induce larger changes in behaviour than one would expect based on the monetary value of the points.

In Japan, in some cities another possibility arises through the provision of transport services by private rail operators. These operators provide bus and rail services and at the same time run supermarkets and department stores. Therefore, point systems might be used to not only encourage travel by public transport in general but to encourage travel by the companies public transport system. This is likely to encourage shopping in their stores also as these are located along the companies' railway line. Furthermore, it is then easy to implement the exchange of the points, similar to credit card points, for shopping.

This chapter provides an example of a private rail operator based in Shizuoka, Japan where the smart card can be used for shopping in supermarkets also. The smart card of this operator includes a point system, similar to air mileage programs.

## 2. MULTIPURPOSE SMART CARDS

The most well-known smart card that can be used for multiple purposes is probably Hong Kong's Octopus card. It can be used for a range of stores including convenience stores, Starbucks or to pay at McDonald's in Hong Kong. Currently, the first generation smart card is replaced by a second generation card that also allows online payments (Octopus, 2015). Similar schemes of multipurpose smart cards are also operated in several other cities around the world. Following the review of Pelletier (2011) Singapore should be mentioned where smart cards are used for a range of services so that they are a common payment mode also by non-transit participants (Smart Card Alliance 2009). Further noteworthy is the T-Money system in South Korea (T-money 2015, Asia IC Card Forum 2012), which can be used for riding taxis, at coin lockers, for public phones and public facilities such as museums.

Regarding impact on travel behaviour, smart cards have certainly made public transport usage more convenient and in large cities most travellers use it, though it seems difficult to show how many new customers the smart card has attracted or whether existing customers travel more nowadays due to the smart card. McDonald (2000) reports that bonus point systems for smart card usage so far appear to have the main purpose to encourage people to sign up for the card. Those smart cards are not much used directly for demand management. This appears still largely true also sixteen years later.

Understanding the potential of bonus points for increasing public transport demand is the topic of the remainder of this chapter. The following two sections offer an introduction to Shizuoka, the Shizutetsu group and the smart card it provides. Section 5 then discusses a stated preference (SP) survey to understand whether travellers would change their behaviour depending on the point system. Section 6 concludes this chapter.

## 3. CASE STUDY AREA AND SMART CARD DATA OVERVIEW

### 3.1 Shizuoka and Shizutetsu

The case study is Shizuoka city in Japan with a population of 703,937 (as of Oct. 2014). Shizuoka is situated on the main Shinkansen line between Tokyo and Osaka. Within Shizuoka, the "Shizutetsu-group" has emerged as a rail operator, bus service provider and a retailer. The Shizutetsu railway is used by about 10 million passengers per year covering 11 km between 15 stations connecting from Shin-Shizuoka station to Shin-Shimizu station. Shizutetsu buses are used by about 30 million passengers per year mainly in the centre of Shizuoka City. The supermarket chain called "Shizutetsu store" has a total of 33 stores of which 19 stores are in Shizuoka city and 14 stores in the surrounding area. Many of these stores are situated near a railway station or bus stop. In addition to the supermarkets there is also a department store (shopping mall) close to Shin-Shizuoka Station. The same smart card could be used for rail usage, bus usage as well as shopping in any of the 33 supermarkets as well as the department store. Furthermore, Shizutetsu group is also operating highway buses, taxis, a cable car and is active in the car rental, building management and other businesses.

## 3.2 Multipurpose Smart Card "LuLuCA"

Regardless of being a public corporation or a private company, most Japanese transport operators, especially railways, have spread their business to include subsidiaries that benefit from the increasing people flow and with it land value growth brought by their transport services. As is the case for Shizutetsu, most rail operators also own shops in the station, supermarkets or rent out shopping and office spaces.

Since smart cards are further often issued by the transport operators themselves, this brings the possibility to use the smart card as a "common currency" for all business branches of the operator. For example, the East Japan Railway Company Group, one of the largest operators in Japan, owns convenience stores called "kiosk" in stations, at which it is possible to pay for food and other articles by the smart card. The JR East smart card can though not yet be used for larger shopping in supermarkets or departments of the store. This is possibly because a credit function is not (yet) available on this smart card.

Shizutetsu group appears to be one step ahead in this respect. The smart card introduced by Shizutetsu is called "LuLuCa". There are four types of LuLuCa depending on whether the cards have a credit or only debit function and further depending on which range of Shizutetsu services they can be used for. With two types of cards only shopping use is possible but not public transportation usage. In the following we refer to these as "shopping only card". Table 1 shows the four types of LuLuCa cards.

**Table 1.** LuLuCa types and spread

| Card Type | | Can Pay for Shizutetsu Public Transport | Can Pay for Shopping in Shizutetsu Group Stores | Number of Cardholders |
|---|---|---|---|---|
|  | LuLuCa Point | — | — | 390,767 |
|  | LuLuCa Pasar | ○ | — | 205,203 |
|  | LuLuCa Plus with Credit Function | ○ | ○ | 29,567 |
|  | LuLuCa Paleta with Credit Function | — | ○ | 16,042 |

Each smart card has a unique ID and is assumed to be used by the same individual over the analysis period. On the two card types with credit function personal information such as home address, age, gender is also provided at the registration.

When LuLuCa was first introduced, users could get an incentive to use the smart card by rewarding them with an extra 100 Yen (about 0.9US $) when the card was charged with 1,000 Yen. That is, charging the card allowed the users to pay for services worth 1,100 Yen. However, in March 2014 a new point service was launched. Since, then the passengers using LuLuCa to pay for public transport get instead 10% of the ride fare as points (except for season cardholders). Further, when people use LuLuCa for shopping they get 1 point for every 108 Yen paid in cash in one of the Shizutetsu supermarkets. (108 Yen and not 100 Yen because 8% VAT is not considered for the points rule). In addition, there are some cases of more points being granted by the stores for products on offer. Whenever a customer/traveller collected 500 points these are exchangeable for vouchers worth 500 Yen that can be used either for travel on Shizutetsu services or for shopping in Shizutetsu-group owned stores.

Table 2 summarizes the data that can be obtained from LuLuCa, distinguished by public transport, shopping related data and personal information. All data include the unique ID, which makes it possible to analyse the three sets together. We note that public transport data includes the fare which allows us to also distinguish commuter pass holders as they are not charged for single trips.

**Table 2.** Data collected by LuLuCa

| Data Type | Data Item |
|---|---|
| Public Transport (Train and Bus) | Card ID (unique value), usage date (year/month/day), usage time when getting on or off the bus or entering/exiting the rail station. Usage fare. |
| Shopping | Card ID (unique value), usage date (year/month/day), usage time of shopping, shop code, purchase amount. |
| Personal Attributes | card ID (unique value), age, gender, home address, family type, income range (only credit card). |

## 4. OVERVIEW OF COLLECTED DATA

Table 3 shows the result of aggregating the number of public transport usage and shopping. The total number of bus trips is about 3 times that of train trips, because the target area has a widespread bus network and only a single train line. LuLuCa is further even more used for shopping than for public transport. Considering that some cards could be used for public transport only, we find though that there are less shopping transactions per cardholder. Therefore, one can conclude that many daily users of Shizutetsu buses do not visit their stores or supermarkets during their journeys. (In general, in Japan the supermarket shopping frequency is probably higher than in most other developed countries, as Japanese tend to buy fresh food several times per week.) Changing this situation is one of the motivations for Shizutetsu to conduct this study. Figure 1 reinforces that further there is an even larger pool of shopping customers who do not use public transport. Obviously, Shizutetsu would like to attract some of these people to both shop in their stores and to use their public transport.

**Table 3.** Aggregate LuLuCa usage statistics in 2014

| | | Total Usage | Number of Cardholders | Average Usage per Card |
|---|---|---|---|---|
| Transit | Train | 6,555,750 | 74,472 | 88.0 |
| | Bus | 18,500,922 | 157,907 | 117.2 |
| Shopping | | 25,079,953 | 426,050 | 58.9 |

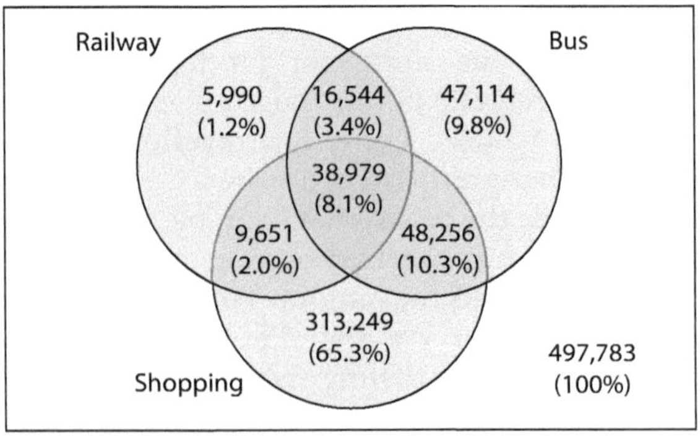


**Fig. 1.** The usage pattern of public transport and shopping

Figure 2 shows the usage pattern of cardholders by gender and age. Overall, women use LuLuCa more since they go shopping more to Shizutetsu supermarkets. Men though use Shizutetsu more for commuting. Comparison of LuLuCa usage by age further shows that there are no significant differences among people in their working ages. Older people use LuLuCa much more for shopping only. This possibly shows an opportunity that improvements in the service could encourage this age group to remain public transport customers.

## 5. STATED PREFERENCE SURVEY ON SENSITIVITY TO POINT SYSTEM

### 5.1. Survey Structure and Hypotheses

The case study aims to understand better sensitivity of behaviour to potential changes in the point system. In particular, it is analysed whether changes in the point system will impact users' mode choice and shopping frequency. With the cooperation of the rail operator, therefore a stated preference (SP) survey is implemented among the card users. Subjects are selected based on their public transport usage and shopping record. In the SP survey, the regular points that a user can earn by using public transport vary as well as details of the "bonus point lottery". Both the values of the bonus points as well as the potential to win are varied. Through ordered regression models it is then analysed which customers are likely to decrease or increase their travel frequency and which customers are not influenced by the point system. The authors are further in the fortunate position that socio-demographics recorded on the smart card such as gender, age and home location can be used as control variables.

A number of potential point schemes are tested with the goal to increase public transport usage and, important for Shizutetsu, to increase shopping in their stores. For this purpose a number of schemes are designed in which extra points are earned if passengers use public transport service and go shopping on the same day. It is expected that such schemes might increase "planned shopping" as well as "incidental shopping".

The questionnaire consists of two main parts. In the first part the respondents are asked about their knowledge of the current point system as well as their travel and shopping behaviour. Furthermore, the travel behaviour of the respondent can be observed through linking the survey respondent with their smart card records via the smart card ID. To get information about their other transport usage the respondents are also asked about their usage of private cars as driver and passenger. In the second part of the survey users are asked about their likely behaviour if a new point scheme would be introduced. Thus, the survey is designed with the following hypotheses in mind:

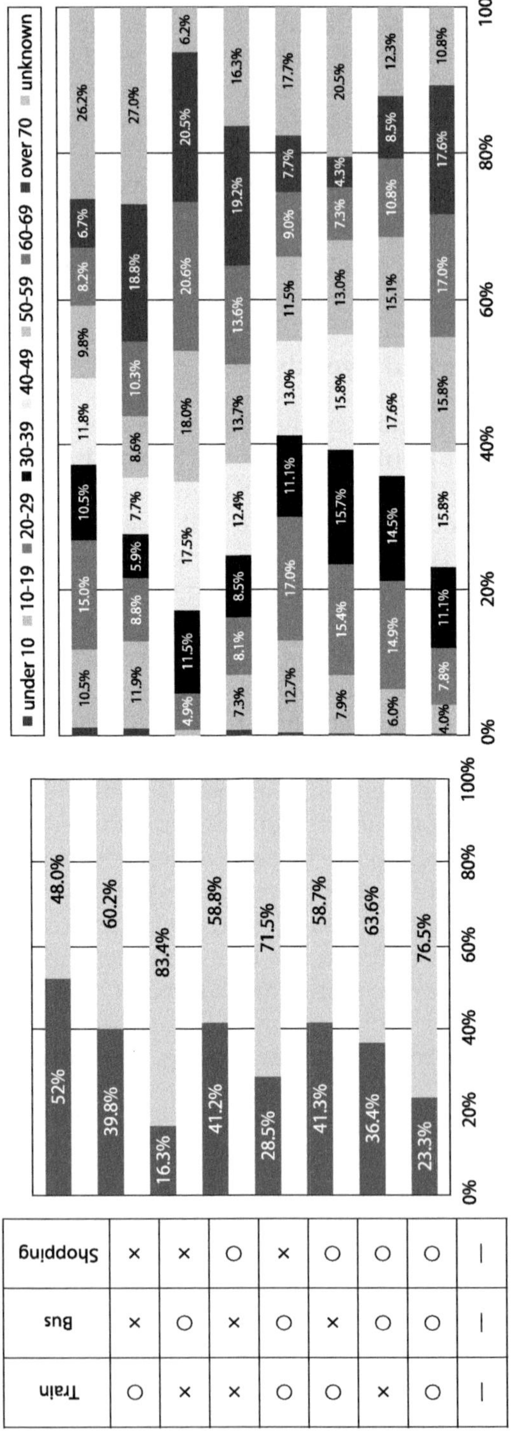


**Fig. 2.** Usage pattern of cardholders by gender and age

***Hypothesis 1***: The behaviour of low frequency public transport users is more sensitive to the point service, compared to card holders who do not use public transport at all and those who use it very frequently.

Our rationale is that a point service is unlikely to persuade those who are not interested in using public transport at all. Further those who rely on public transport already are unlikely to make more trips just because of the point system. Instead, occasional, infrequent users who are in general willing to use public transport are likely those who might increase their bus or train journeys if provided more incentives.

***Hypothesis 2***: A point service combined with a lottery has the potential to increase the usage of public transport, while reducing the burden on the operators by appropriately setting the winning chance.

Lotteries are commonly applied for many existing point systems. The idea could be linked to prospect theory, that is, users might be disproportionately attracted to potential large but risky gains.

**Table 4.** SP settings for certain points and for bonus points

| | Level 1 | Level 2 | Level 3 |
|---|---|---|---|
| Certain points | 10% of fare | 5% of fare | 2.5% of fare |
| Bonus points winning probability | 90% | 50% | 10% |
| Expected bonus point | 0 | 20.5 | 45 |

To verify the above hypotheses, three factors in the SP survey vary between three levels respectively as shown in Table 4. "Certain points" means the number of points given uniformly according to the paid fare when boarding a train or bus. In the current point service 10% of the fare is given. In addition, the impact if only 5% or 2.5% of the fare is being "returned" as points is tested.

Only those customers who use public transport and go for shopping in the same Shizutetsu supermarkets enter the bonus point lottery. Expected bonus points describe the amount of bonus points given divided by the winning chance. For example, in case of 90% winning probability and 45 expected bonus points, 90 out of 100 people get 50 bonus points whereas 10 receive none. Zero bonus points are added as a case to test the "base case" of no lottery. These factors were allocated using an experimental design (L9 orthogonal array).

Table 5 shows the information that was actually presented to the respondents. For an SP it is desirable for the order of the 9 patterns posed to respondents to be perfectly random. This is approximated by asking respondents 5 out of 9 patterns that are chosen according to the smart card number. For each pattern, then the respondents are asked if they are likely to {increase, not change, decrease} the number of times they travel by public transport and go shopping on the same day.

**Table 5.** Patterns presented to survey respondents

| Point scheme ID | Certain points | Bonus points winning Probability | Number of bonus point if lucky |
|---|---|---|---|
| 1 | 10% of fare | 0% | 0pt |
| 2 | 10% of fare | 50% | 45pt |
| 3 | 10% of fare | 10% | 450pt |
| 4 | 5% of fare | 90% | 25pt |
| 5 | 5% of fare | 50% | 90pt |
| 6 | 5% of fare | 0% | 0pt |
| 7 | 2.5% of fare | 90% | 50pt |
| 8 | 2.5% of fare | 0% | 0pt |
| 9 | 2.5% of fare | 10% | 225pt |

In addition respondents are asked to describe in words how much their travel + shopping frequency might change. The survey was conduct as a mail return survey randomly selecting LuLuCa cardholders. In total, 5,891 survey forms were sent out and 1,302 completed surveys are obtained which is equal to a response rate of 22.1%.

## 5.2 Descriptive Survey Results

Figures 3 and 4 illustrate the knowledge about the current point service among the cardholders. There are 435 respondents who have invalid cards that cannot be used for public transport as shown in Figure 3. It was found that 38% are not aware of the point scheme, which is not surprising since about 52% of the cardholders do not use it for public transport. 49% of those who do not use a card that allows for public transport usage do not know about the point service. It therefore suggests that a simple measure for Shizutetsu to potentially increase public transport usage is to start a campaign to raise awareness of the point system. In contrast to the public transport usage point system, the shopping point system is widely known (more than 90% of cardholders), as illustrated in Figure 4.

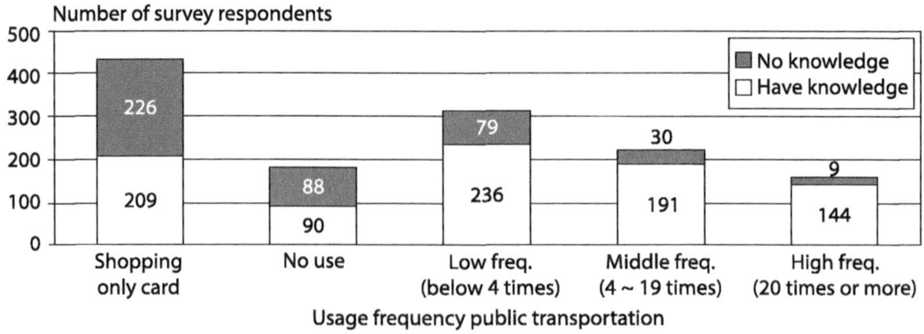


**Fig. 3.** Knowledge about the PT usage point service

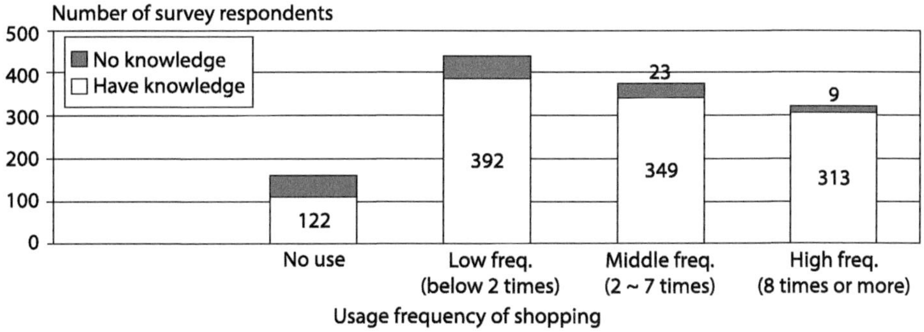


**Fig. 4.** Knowledge about the shopping point service

Figure 5 shows the percentage of subjects who answered "increase", "no change", or "decrease" concerning the usage of public transport and shopping on the same day. Note that the first scheme is in fact the currently operating scheme. Still nearly 30% of customers answered that they will increase their usage frequency under this point scheme and 5.6% answer that they will decrease the use. Therefore, these numbers might partly show the missing knowledge of the point scheme but also the scale of a typical SP response bias where the respondents tend to express "good intentions". These biases should hence be taken into account when interpreting our subsequent results. It is found that 568 respondents replied that they will not change their behaviour independent of the point scheme and this group is referred to in the following as the "no change group". The focus of the subsequent analysis is on the 734 users who indicated that their usage might vary based on the point scheme. Among these "changers" the expected tendency that more rewarding point services lead respondents to reply that they will more likely use public transport and go shopping is found.

## 5.3 OLM and MNL Analysis

Ordered logit multinomial (OLM) choice models are formulated with intended change in usage frequency as dependent variable. As discussed above a number of explanatory variables could be obtained. Firstly, the survey respondents who might change their behaviour are divided into 12 groups according to their current usage of the smart card for public transport and shopping as shown in Table 6. Additional explanatory variables and interaction terms tested are listed in Table 7.

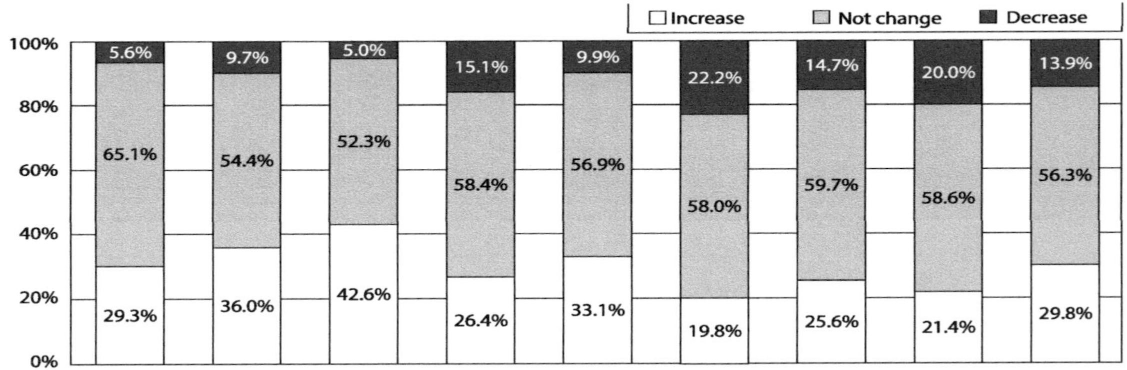


| Scenario number | 1 | 2 | 3 | 4 | 5 | 6 | 7 | 8 | 9 |
|---|---|---|---|---|---|---|---|---|---|
| Certain points | 10% of fare | 10% of fare | 10% of fare | 5% of fare | 5% of fare | 5% of fare | 2.5% of fare | 2.5% of fare | 2.5% of fare |
| Bonus points winning probability | 0% | 50% | 10% | 90% | 50% | 0% | 90% | 0% | 10% |
| Expected bonus points | 0 | 22.5 | 45 | 22.5 | 45 | 0 | 45 | 0 | 22.5 |
| Number of bonus points | 0pt | 45pt | 450pt | 25pt | 90pt | 0pt | 45pt | 0pt | 225pt |

**Fig. 5.** The scenarios and user responses

**Table 6.** SP respondents grouped by frequency of shopping and public transport usage

| | | LuLuCa Shopping per Month | | | Total |
|---|---|---|---|---|---|
| | | None (zero times) | Occasionally (1-7 times) | Frequent (8+ times) | |
| LuLuCa PT Usage per Month | LuLuCa not registered for PT usage | 74 | 153 | 16 | 243 |
| | None (zero times) | 62 | 92 | 30 | 284 |
| | Occasionally (1-19 times) | 95 | 80 | 43 | 218 |
| | Frequent (20+ times) | 20 | 56 | 13 | 89 |
| | Total | 251 | 381 | 102 | 734 |

**Table 7.** Explanatory variables for OLM and MNL models

| Name of Explanatory Variables | Description |
|---|---|
| Frequency of current PT × Frequency of current shopping | Interaction of the variables shown in Table 6. These twelve groups are further interacted with certain points and expected bonus points in the model. |
| Certain points | Certain points to be obtained when cardholder pays for PT. |
| Expected bonus point | Expected bonus point to be obtained when the cardholder uses PT and goes shopping on the same day. |
| Winning probability | Winning probability for bonus points. |
| Average PT usage per month | Mean times of train and bus usage by LuLuCa from December 2013 to November 2014. |
| Expected decrease in certain points | Reduction in points given the current public transport usage and the one stated in the scenario. |
| Amount of shopping per month | Amount of shopping per month with LuLuCa between December 2013 and November 2014. |
| Private car usage dummy | 1 for cardholders using their private car more than two days per week. |
| Living with family dummy | 1 for cardholders living with their family. |
| Living alone dummy | 1 if cardholders live by themselves. |
| Near station dummy | 1 for cardholders who live within 700 m (direct distance) from a station. |
| Near bus stop dummy | 1 for cardholders who live within 200 m (direct distance) from a bus stop. |
| Near supermarket dummy | 1 for cardholders who live within 500 m (direct distance) from a Shizutetsu supermarket. |
| No train usage dummy | 1 if the card has not been used for travelling by train between December 2013 and November 2014. |
| No bus usage dummy | 1 if the card has not been used for travelling by bus. |
| No shopping usage dummy | 1 if the card has not been used for shopping. |
| PT point knowledge dummy | 1 if the respondent knows about the point service for PT usage. |
| Store point knowledge dummy | 1 if the respondent knows about the point service for shopping. |

First, an ordered logit model is estimated where the dependent variable is whether the respondent is planning to increase, not change or decrease his frequency of using PT and go to a Shizutetsu store for shopping on the same day. Then a multinomial logit model (MNL) is estimated again with same three response categories as dependent variable. Table 6 shows the estimated results of both models.

As expected, parameters related to points and winning probability mostly take a positive value. Hence, as the points and winning probability become larger, respondents are more likely to choose "increase". Parameters related to the fixed points given are significant for eight out of twelve groups. In particular, it is found that those who already at least occasionally shop and use public transport state that they would increase their shopping and public transport usage. Parameters for current non-public transport users are instead not or barely significant. This suggests that by changing the point system it is difficult to attract new travellers/customers but the existing travellers/customers could be persuaded to travel and shop more often if they promise more points. Similarly, it is found that lottery type bonus points might attract existing customers but not people who now either do not shop or do not use Shizutetsu transport services at all. Our further parameter estimates suggest that in particular the current PT usage frequency determines whether the point system would have an impact.

Through dummy variables, it was found further that the family situation and whether a person is living near a station are significant factors. Those living near a station are less likely being influenced by the point system, possibly due to the fact that they also depend more on the rail and bus services and are captive users anyway.

The purpose of the MNL model is to test whether there are different effects determining "increase" and "decrease" compared to the reference group "no change". It is found that reduction in certain points is more significant to explain why some person groups might decrease their public transport and shopping usage, whereas lottery points are more significant to explain why some people would increase their shopping and public transport usage. This suggests that any reduction in certain points should be considered carefully as it would possibly affect the behaviour of existing customers. Lottery points appear risky for the operator also as they might attract new customers, but it is not clear if there are significant increases in sales/ridership. Through the MNL model it is further found the car usage dummy is positive significant for "no change". This once more reinforces the observations that point systems seemed of limited effect to attract new customers.

**Table 8.** OLM and MNL modelling results

| | | | | Ordered Logit Model | | Multinomial Logit Model "increase" | | Multinomial Logit Model "no change" | | Multinomial Logit Model "decrease" | |
|---|---|---|---|---|---|---|---|---|---|---|---|
| | | | | Parameter | (t-stat) | Parameter | (t-stat) | Parameter | (t-stat) | Parameter | (t-stat) |
| OLM: Threshold 1/MNL: Constant "decrease" | | | | −0.62 | (−2.88) ** | | | | | 0.30 | (−2.15) ** |
| OLM: Threshold 2/MNL: Constant "no change" | | | | 2.25 | (−10.40) ** | | | 0.87 | (5.48) ** | | |
| Certain points | shopping only card | × | No shopping | 0.52 | (1.35) | 0.09 | (2.06) * | | | 0.02 | (0.31) |
| | shopping only card | × | Occasional shopping | 0.75 | (5.41) ** | 0.06 | (4.05) ** | | | −0.07 | (−3.02) ** |
| | shopping only card | × | Frequent shopping | 0.76 | (4.27) ** | 0.07 | (3.49) ** | | | −0.07 | (−2.39) |
| | No PT | × | No shopping | 0.31 | (0.77) | 0.04 | (0.87) | | | −0.02 | (−0.42) |
| | No PT | × | Occasional shopping | 0.33 | (1.55) | 0.00 | (0.04) | | | −0.09 | (−2.29) * |
| | No PT | × | Frequent shopping | 0.58 | (2.47) * | 0.05 | (1.83) | | | −0.08 | (−2.04) * |
| | Occasional PT | × | No shopping | 1.10 | (4.04) ** | 0.01 | (0.28) | | | −0.17 | (−3.99) ** |
| | Occasional PT | × | Occasional shopping | 1.08 | (6.22) ** | 0.03 | (2.05) * | | | −0.10 | (−4.79) ** |
| | Occasional PT | × | Frequent shopping | 0.99 | (4.50) ** | 0.04 | (1.67) | | | −0.08 | (−2.61) ** |
| | Frequent PT | × | No shopping | 0.31 | (0.80) | −0.05 | (−1.09) | | | −0.09 | (−1.80) |
| | Frequent PT | × | Occasional shopping | 1.12 | (4.58) ** | 0.04 | (1.62) | | | −0.13 | (−3.91) ** |
| | Frequent PT | × | Frequent shopping | 0.95 | (2.19) * | 0.03 | (0.66) | | | −0.12 | (−1.79) |
| Expected bonus point divided by 100 | shopping only card | × | No shopping | 0.17 | (1.41) | 0.03 | (0.24) | | | −0.36 | (−1.53) |
| | shopping only card | × | Occasional shopping | 0.07 | (1.83) | 0.05 | (1.22) | | | −0.08 | (−1.12) |
| | shopping only card | × | Frequent shopping | 0.08 | (1.44) | 0.06 | (1.00) | | | −0.11 | (−0.96) |
| | No PT | × | No shopping | 0.19 | (1.41) | 0.15 | (0.99) | | | −0.17 | (−0.72) |
| | No PT | × | Occasional shopping | 0.22 | (2.97) ** | 0.17 | (2.06) * | | | -0.58 | (-2.31) * |
| | No PT | × | Frequent shopping | 0.14 | (1.72) | 0.12 | (1.38) | | | −0.17 | (−0.96) |
| | Occasional PT | × | No shopping | 0.23 | (3.63) ** | 0.25 | (3.47) ** | | | −0.20 | (−1.22) |

*Contd...*

| | | | | | | | | | | | | | | |
|---|---|---|---|---|---|---|---|---|---|---|---|---|---|---|
| Expected bonus point divided by 100 | Occasional PT × Occasional shopping | 0.22 | (6.31) | ** | 0.20 | (5.31) | ** | | | | –0.19 | (–2.49) | * |
| | Occasional PT × Frequent shopping | 0.13 | (2.15) | * | 0.11 | (1.68) | | | | | –0.13 | (–1.15) | |
| | Frequent PT × No shopping | 0.09 | (0.85) | | 0.07 | (0.46) | | | | | –0.16 | (–0.86) | |
| | Frequent PT × Occasional shopping | 0.12 | (2.05) | | 0.11 | (1.70) | | | | | –0.09 | (–0.78) | |
| | Frequent PT × Frequent shopping | -0.01 | (-0.07) | | –0.10 | (–0.62) | | | | | –0.23 | (–0.86) | |
| Winning probability | | 0.29 | (4.08) | ** | 0.24 | (2.81) | ** | | | | –0.18 | (–1.75) | |
| Av. PT usage per month $\times 10^{-1}$ | | 0.08 | (2.36) | * | | | | -0.07 | (-2.18) | * | | | |
| Expected decrease in certain points | | 0.01 | (1.41) | | | | | | | | 0.01 | (2.85) | ** |
| Amount of shopping per month in supermarket $\times 10^{-4}$ | | 0.01 | (0.24) | | | | | 0.05 | (2.33) | * | | | |
| Private car usage dummy | | 0.09 | (1.53) | | | | | 0.18 | (3.00) | ** | | | |
| Living with family dummy | | 0.17 | (2.83) | ** | 0.22 | (3.59) | ** | | | | | | |
| Living alone dummy | | 0.09 | (0.93) | | | | | | | | 0.08 | (-0.68) | |
| Near station dummy | | –0.19 | (–2.70) | ** | -0.16 | (-2.06) | * | | | | | | |
| Near bus stop dummy | | –0.08 | (–1.59) | | –0.06 | (–1.13) | | | | | | | |
| Near supermarket dummy | | 0.09 | (1.47) | | 0.12 | (1.69) | | | | | | | |
| No train usage dummy | | 0.13 | (1.67) | | | | | 0.08 | (0.52) | | | | |
| No bus usage dummy | | 0.41 | (3.24) | ** | | | | 0.19 | (2.70) | ** | | | |
| No shopping usage dummy | | 0.01 | (0.06) | | | | | 0.08 | (0.52) | | | | |
| PT point knowledge dummy | | –0.03 | (–0.44) | | | | | –0.01 | (–0.14) | | | | |
| Store point knowledge dummy | | 0.09 | (1.02) | | | | | 0.16 | (1.80) | | | | |
| Number of observations | | 6,606 | | | | | | Number of observations | | | 6,606 | | |
| Log likelihood at the initial value | | –7,704 | | | | | | Log likelihood at the initial value | | | –7,257 | | |
| Log likelihood at convergence | | –6,083 | | | | | | Log likelihood at convergence | | | –6,042 | | |
| Likelihood ratio index | | 0.21 | | | | | | Likelihood ratio index | | | 0.17 | | |
| Likelihood ratio index adjusted for the degrees of freedom | | 0.21 | | | | | | Likelihood ratio index adjusted for the degrees of freedom | | | 0.16 | | |

**:1% significance *:5% significance

## 6. CONCLUSION

After a general discussion on multipurpose smart cards a combined point service for public transport use and shopping in Shizuoka, Japan is evaluated. The fare system initially introduced by Shizutetsu had been very generous in providing travellers with a 10% discount for prepaid cards. This system was then changed in 2014 to a system where users get points according to the amount of fare paid. The implicit amount of discount has not been changed as still 10% of the fare is returned to the users in the form of points. Nevertheless, the use of public transport has decreased. Partly because of this and partly because the point system is still generous compared to points earned by shopping, Shizutetsu is now considering alternative point schemes.

With the effect of an SP survey the possible user reactions are evaluated. The effect of reducing the fixed points given to users are tested as well as the potential impact of a lottery where users can win points if they travel by public transport and shop on the same day. This requirement of having to shop at Shizutetsu stores and use public transport on the same day to enter the lottery is obviously of interest to Shizutetsu but one might argue can also be interpreted as a sustainable transport policy in aiming to shift the mode choice especially for shopping trips.

It is found that there is a substantial number of users (around 56%) who will not change their travel and shopping behaviour in response to changes in the point system. However, through OLM and MNL modelling it is also found that the users who are already using public transport or shopping are likely to reconsider their behaviour and might in future increase their shopping frequency (if they have been frequent public transport users already) or their public transport usage (if they have been regular Shizutetsu supermarket customers already). It is further found that especially those with LuLuCa cards that are not currently registered for PT usage might change their card and hence start using the scheme. Another important finding is that small decreases in the point system might not have a significant effect especially if compensated with a lottery system.

Clearly the current results cannot be taken as direct predictions on how much the demand will change due to potential biases in the SP responses. However, it is believed that the general tendencies in change of behaviour obtained for the different groups would pertain. As one limitation it should be mentioned that the study could consider cardholders only so that it does not include the effects of any customers currently not possessing a LuLuCa. However, especially for this reason the results regarding the users who currently have LuLuCa cards are not eligible for using public transport appears promising.

In general, it is hoped that this study might trigger more research on promoting change in behaviour through new pricing options available to operators through smart card data, in particular through increasingly available multipurpose smart cards. This appears to be still an under-researched area.

## REFERENCES


Asia IC Card Forum: WBS AFC Standard SW Solution Development (KOREA). 2012. <http://www.asiaiccardforum.net/news/02-01-2012-01/data/AFC-KSCC-Korea.pdf> (visited on 05.10.2015).

Davis, D. 1999. *Ten Projects that Shaped the Smart Card World Technology*. December.

EZ-Link "*Continue your Journey Everyday with Us*", (2015).<http://home.ezlink.com.sg/> (visited on 05.10.2015).

McDonald, Noreen. 2003 *Multipurpose Smart Cards in Transportation: Benefits and Barriers to Use,,* Spring 630. Available from <http://www.uctc.net/research/papers/630.pdf>. Accessed Nov. 2015.

Michael, L. 2004. The influence of loyalty programs and short-term promotions on customer retention. *Journal of Marketing Research*, 41, pp. 281-292.

Morency, C., Trepanier, M. and Agard, B. 2006. Analysing the Variability of Transit Users Behaviour with Smart Card Data. *The 9th International IEEE Conference on Intelligent Transportation Systems*. ITSC, Toronto, Canada, September 17-20. pp. 44-46.

Morikawa, T. 2008. Eco-Transport Cities Using ITS. *IATSS Research*. 32(1), pp. 26-31.

Meadowcroft, P. 2005. Hong Kong raises the bar in smart card innovation. *Card Technology Today* (January), pp. 12-13.

OCTOPAS. 2015. *Making Everyday Life Easier*. <http://www.octopus.com.hk/home/en/index.html> (visited on 05.10.2015).

Pelletier, M. -P., Trepanier, M. and Morency, C. (2011). Smart Card Data Use in Public Transit: A Literature Review. *Transportation Research Part C*, 19, pp. 557-568.

Sato, H., Kurauchi, S., Morikawa, T. and Yamamoto, T. 2006. A Study on Differences between Travel Eco-Point System and Fare Reduction of Public Transportation -An Empirical Analysis based on Mental Accounting Theory. *Papers of Research Meeting on Civil Engineering Planning*, Vol. 34 (in Japanese).

T-money. 2015. Korea Smart Card Co., Ltd. <http://eng.t-money.co.kr/> (visited on 05.10.2015).

Taylor, G.A. and Neslin, S.A. 2005. The current and future sales impact of a retail frequency reward program. *Journal of Retailing*. 81(4), pp. 293-305.


## AUTHOR BIOGRAPHY

**Toshiyuki Nakamura** is an Assistant Professor in the Graduate School of Engineering at Kyoto University. Toshiyuki has been engaged in various research on travel behaviour and road project evaluation, in particular bus users, using the "PASMO" smart card that is operated in the Tokyo metropolitan area. He has been a member of IBS, Japan transport planning consultancy before joining Kyoto University in 2011. Nowadays he has engaged in several researched with Shizutetsu regarding the use of the "LuLuCa" data described in this chapter.

**Nobuhiro Uno** is an Associate Professor in the Graduate School of Management at Kyoto University. His research interests include travel behaviour under provision of information, traffic control for urban expressway and traffic conflict analysis of vehicular behaviour using video image data. In addition, he and his colleague evaluated a new inter-city expressway using ETC (Electronic Toll Collection) card data to measure behavioural changes and travel time reliability.

**Jan-Dirk Schmöcker** is an Associate Professor in the Graduate School of Engineering at Kyoto University. Jan-Dirk's research interests cover modelling of network flows as well as data driven analysis of travel behaviour. He has published work related to analysis of London's Oyster card data and has been involved in studies using smart card data from Japan. His current research interest focuses on understanding and modelling the gradual changes in behaviour over time of individuals.

**Takenori Iwamoto** is a manager at Shizuoka Railway Co. Ltd. He has been involvement in the initial development of the "LuLuCa" smart card in 2006. Since then he is continuously utilizing accumulated information from the card and is making use of the data for the marketing of Shizuoka Railway and other branches of the company. Currently, he is furthermore using smart card data to write a PhD thesis related to demand prediction and topics discussed in this chapter.

Chapter 8

# Using Smart Card Data for Agent–Based Transport Simulation

**P.J. Fourie[1,*], Alex Erath[2], S.A. Ordonez[3], A. Chakirov[4] and K.W. Axhausen[5]**

**ABSTRACT**

The disaggregate nature of transit smart card data is congruent with the travel demand specification as used by agent-based approaches to transport modelling. Using a full day of public transport smart card transactions recorded in Singapore, we develop an approach to transform the smart card data into both transport supply and demand, while simultaneously eliminating the need to simulate the interaction between cars and buses. In order to produce realistic travel times for buses, a regression model of bus speed between stops was estimated, that is dependent both on the level of demand and network topology. The simulation includes a model of bus dwell time at stops that is dependent on the ridership of the bus and its configuration. It allows us to simplify the supply network dramatically with only one link between bus stop combinations and another link at the stop for buses to queue to perform dwell operations. These modifications, along with a simplified mobility simulation, dramatically improves simulation times, ensuring useable results in an hour. In addition, our modelling framework is highly adaptable and requires only limited efforts to be applied to other public transport systems in cities where similar data streams are available.

## 1. INTRODUCTION

It is widely agreed that provision of attractive public transport services is of central importance for the sustainable development of cities, as it outperforms individual motorized transports in terms of cost, environmental impact and social equity. To plan and design efficient urban public

[1-5]Future Cities Laboratory, Singapore ETH Centre, #06-01 1 CREATE Way, Singapore, 138602.
* Corresponding author

transport service provision, municipal planning organizations (MPOs) and service operators usually develop transport demand models. The models currently used in practice operate on the principle of modelling trip flows between geographical zones and hence are subject to aggregation over the horizons of time and space. However, current urban transportation problems, such as congestion and service reliability, are of an inherently dynamic nature. This is particularly the case for public transport as overcrowding, schedule reliability and bus bunching are inherently dynamic phenomena observed in many cities all over the world.

To address the shortcomings of aggregate methods, large-scale, agent-based transport demand simulation models have been developed that preserve full temporal dynamics as well as disaggregated information on individuals through the entire modelling and simulation process. Software packages such as MATSim (Balmer et al. 2009) or TRANSIMS (Smith et al. 1995) are designed to dynamically simulate transport demand and supply for millions of agents over an entire day at the temporal resolution of a second. These models take an activity-based approach, acknowledging that travel demand is the result of the need to perform activities in different points in space and time. Entities in the simulation have a one-to-one correspondence with their real-world equivalents, therefore an agent in the simulation represents a single commuter in the physical system. Similarly, private and public transportation vehicles have equivalent entities in the simulation. Dynamic phenomena such as congestion and bus bunching emerge from interactions between all participating agents in the simulation.

While many cities recognise the potential of agent-based and activity-based approaches, these methods have come under much criticism for being exceptionally data hungry, with finely-grained information needed at all stages of the travel demand modelling process. These typically include a detailed synthetic population describing travel demand as function of various household and personal demographic attributes; as well as the modelling of the transport supply system, i.e., the transportation network, vehicle fleet, public transport schedule and activity facility capacities serving the demand. Dynamic assignment models are also notoriously difficult to calibrate, as observed traffic volumes and travel times are emergent phenomena resulting from the dynamics in the system. Because these systems work from the bottom up, they require that individual behaviour and interactions be described adequately in the simulation, in order for the full range of dynamic phenomena to emerge as observed in reality. Furthermore, as all participating transportation modes subject to the full range of commuter choice dimensions need to be simulated repeatedly for the system to reach a steady state, simulation times are notoriously long.

In the light of these difficulties, there remains an argument for the use of so-called direct demand models that do not attempt to capture the full gamut of cause and effect as being attempted in the activity-based methods, but instead rely on inferring the reaction of the transportation system to dynamically changing demand, based on observation.

The data produced by automatic fare collection (AFC) systems represents a uniquely appropriate input to a direct demand model. Production of the data records by such systems document public transport ridership patterns in great detail, at the level of individuals with precise spatio-temporal information. Since 2005, authors such as Bagchi and White (2004, 2005) have studied the potential of using this type of data. Pelletier et al. (2011) present a summary on how AFCS data has been used to analyse public transport systems worldwide. While they find that smart card data are used at all three levels of public transport management, i.e. strategic, tactical and operational, their survey uncovers no work so far in using the data to drive a simulation model to predict the performance of alterations to the system.

The aim of this paper is to assess the potential of using AFC system data in an agent-based simulation for the case of Singapore. To this end, a MATSim scenario was created, using smart card transaction data as travel demand input and detailed public transport schedule information and a global positioning system (GPS) navigation network to describe supply. To eliminate the need for simulating the full transport system of public and private vehicles for realistic travel times to emerge from repeated network loading, we derived a model for the speed of buses between public transport stops as a function of localised public transport demand from the smart card data and topographical information contained in the network description. We extend the existing MATSim framework by introducing stochastic terms to describe bus dwell time behaviour at stops and travel time between the stops, which are the main determinants of service reliability in public transport operations. We validate the resulting model against the information contained and derived from the smart card data.

The potential of the approach in predicting the response to alterations in the system is presented by splitting a long existing bus line as a case study. The paper concludes by evaluating the approach's applicability for practice and identifying future research directions.

## 2. USER EQUILIBRIUM AND PUBLIC TRANSPORT IN MATSIM

MATSim is a platform to simulate transport demand and supply interactions allowing for large-scale scenarios where millions of agents represent people interacting. For each agent, a daily activity plan is assigned representing the sequence of activities it has to perform at different times and at different locations within a specific period of time (in general one day). MATSim uses an evolutionary algorithm to reach a steady state. The same day is simulated many times and a fraction of the agents modify their plans after each iteration. There are many ways to modify their plans; they can change the departure time, the travel mode of a sub-tour, or location of a given type of activity, among others. This work focused on the route modification, more specifically for public transport users. The utility of the day is measured for each agent in each iteration using a scoring function that rewards agents for performing activities, while penalizing them for travelling, transferring between transport modes, waiting at transit stops and arriving late for activities, etc. (Charypar and Nagel 2005). Agents save a small number of plans, remembering those that scored well and forgetting the others. Thus, the general score of the population tends to grow until, after hundreds of iterations, the system reaches user equilibrium and the generalized utility cannot be improved any more (Balmer et al. 2009).

MATSim includes a full implementation of public transport (Rieser 2010). On the transportation supply side, the system is represented by stop facilities and transit lines. Several routes can belong to each line. Each of these transit routes holds the information about the sequence of stop facilities with the expected arrival and departure offsets, the sequence of links in the road network a vehicle of this route has to follow and the departure times of all the services of the route. As the links that public transport vehicles have to follow belong to the road network that private vehicles use in the simulation, public transport vehicle travel times are affected by congestion for modes like bus or tram that share the network with private transport, while modes with exclusive networks and precise signaling and control (i.e., rail, subway, monorail) tend to work close to scheduled times.

Another source of deviation from schedule, especially for bus, is the time spent at bus stops allowing passengers to board and alight. This dwelling process could be modelled in two ways in MATSim: the simple approach just calculates the time a vehicle has to stop according to the number of passengers, type of vehicles, number and configuration of vehicle entrances as well as exits and vehicle occupancy, while a more fine-grained approach would simulate a queue agents use to enter the vehicle.

For the bus stop facilities, availability of a bus bay could be specified to indicate whether a bus is obstructing a link for cars during the dwelling process. MATSim also allows the same vehicle be scheduled to perform several services; if it is late, the next service would not be able to start. Thus, the level of detail of the public transport module can simulate phenomena such as early or late services, crowded vehicles, bus or train bunching and long waiting times resulting from service denial to fully loaded vehicles.

## 3. CEPAS

### 3.1 Suitability of Using CEPAS Data to Describe Public Transport Demand

Contactless, stored value smart cards for fare collection have been introduced in Singapore in April 2002 under the name EZ-Link. In 2009, a new standard for electronic payment smart cards called Contactless e-Purse Application (CEPAS) superseded EZ-Link. CEPAS-compliant smart cards could be used island-wide for payment of all modes of public transport, regardless of operator, as well as for minor retail transactions, parking and road toll payment. Though cash payment of single fares at higher rates is still possible, e-payments with CEPAS cards account for 96% of all trips, which makes the data records from CEPAS highly comprehensive and the missing cash paying travellers negligible (Prakasam 2008).

In Singapore, the fare system is distance-based and customers have to tap their CEPAS card on the reading device every time they enter and leave a train station or a bus, or they get charged the maximum amount for that particular service. GPS devices on buses ensure that each transaction has a unique transit stop identifier, as well as the vehicle identification number. Each transaction thus has information on timing and location, and generally most trips contain information on both boarding and alighting transactions; a notable exception is the case of concession cards for schoolchildren, students and senior citizens where the maximum charge is capped at 7.2 km. These users therefore sometimes do not tap out, especially when the bus is full and users want to alight faster.

The completeness of the Singaporean smart card data, both in terms of market penetration and recording of both boarding and alighting locations, distinguishes it from those collected by the majority of other automatic fare collection systems and allows for more detailed assessment of travel behaviour and mobility patterns. In many other countries users do not have to tap out of the bus or tram and the alighting location is thus not recorded, although researchers recently proposed techniques to impute its value in the absence of such information (see Chapter 2 or Munizaga and Palma 2012). Furthermore, as the CEPAS cards are durable and easily rechargeable, people tend to continuously use one single CEPAS card with a unique card ID for all their public transport journeys for substantial periods of time. As the technical setup of the system doesn't allow more than one person to travel on a single CEPAS card, it could be assumed that each unique card ID represents a single person. This enables highly disaggregated analysis of each itinerary and opens new ways for understanding people's travel behaviour over the short as well as longer term scales.

Given the temporal and spatial resolution of the CEPAS data, it is perfectly suited to represent travel demand in a simulation of Singapore's public transport system using MATSim. By combining it with information on supply derived from published schedule information it becomes possible to generate a simplified MATSim scenario. This scenario could be used as a predictive system to evaluate changes in public transport service provision such as the type of buses being used, service frequency and service network.

## 3.2 Combining Agent-based Transport Simulation and CEPAS Data

CEPAS data only describes demand for public transport services. Therefore, we restrict the scope of the MATSim model presented in this study to public transport only and simulates its operation separate from other transport modes.

The scope of analysis is restricted to route choice effects as the simplified scenario covers the public transport system in isolation and no information is available about the trip purposes or socio-demographic background of travellers. Furthermore, the system cannot account for mode choice effects, i.e. passengers switching away from private modes or switching to public transportation due to changes in system performance; neither can it account for so-called induced demand, where changes in the level of performance of the public transportation system result in people performing more or fewer activities because of more or less time opening up in their travel time budgets.

Information of an individual traveller is restricted to a unique card identifier and fare type category, namely, child/student, adult and senior. Furthermore, there is no information about the trip purpose and real trip origin and destination at the level of buildings, but only the public transport stop where the transaction took place. As our approach does not infer the real trip origin, destination or trip purpose, the scope of the analysis cannot include destination choice effects.

To restrict the scope of the simulation to public transport it needs to account for interaction effects with cars resulting in increased travel times. To this end the observed travel times from the smart card data are used to develop a regression model of bus movement between the transit stops. The simulation uses the error term from this regression model to arrive at a stochastic model of travel times between the transit stops that eliminates the need to model a fully detailed network during the simulation. This bus speed model allows one to predict the distribution of travel times during the course of the day for network links that are not in existence now, making system-wide network re-design evaluation possible.

# 4. METHOD

In MATSim, public transport vehicles share roads with other vehicles and dwell operations are modelled in detail. As the aim of this work is to set-up a simulation only using AFC system data, we simplified the MATSim mobility simulation and the transport network, restricting it to public transport vehicles only. Fig. 1 shows the processes we designed and implemented and how these affect a standard MATSim simulation. The next section describes reconstruction of the bus trajectories, generation of a public transit schedule, then generation of public transport trips as MATSim plans (the demand), followed by the simplification of the road network and finally the new mobility simulation model and its constituent sub-models in MATSim.

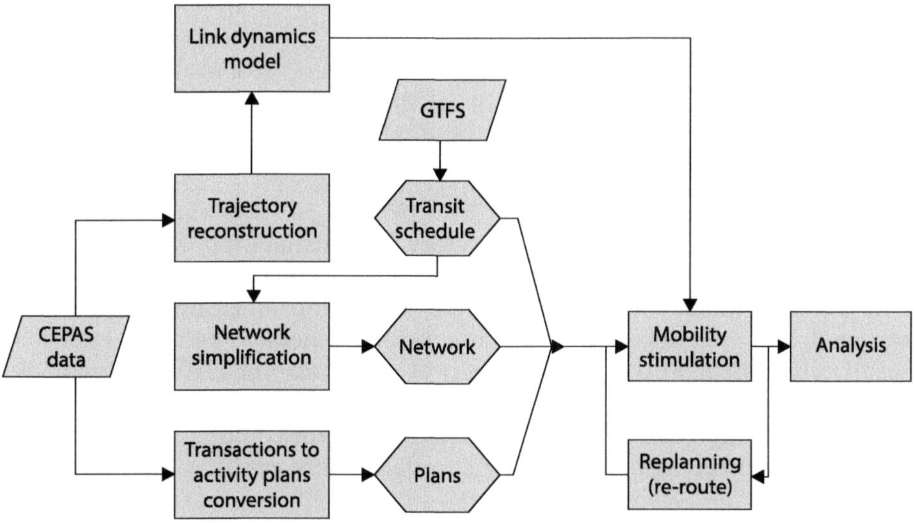


**Fig. 1.** Simplified public transport simulation overview

## 4.1 Reconstruction of Bus Trajectories

Given boarding and alighting transactions of bus users it is possible to estimate the position in space and time of the corresponding buses (their trajectories). For each vehicle ID in the system, by grouping its transactions at each stop into sets that represent bus dwell operations, it is possible to impute the time that it takes for the bus to travel between bus stop locations. From our electronic transit route profile, the exact route between bus stop locations is known and therefore the vehicle's trajectory can be reconstructed once all dwell operations have been identified.

There are a number of challenges in the trajectory reconstruction process:

***Bus stops without transactions***: As boarding and alighting actions might not occur at every stop, the bus can remain "invisible" to the system. A simple interpolation technique was used to estimate the time when the bus reached these stops. For stops that precede or follow the first and last "visible" stops we did not apply extrapolation to estimate the bus arrival times at those stops.

***Early tap-outs and late tap-ins***: As the bus approaches the public transport stop, the GPS system automatically activates the reading device, making it possible to tap out before the bus doors have opened. Furthermore, sometimes passengers have entered the bus but are still fumbling to get their cards out for the reader and the tap-in registers late. As these transactions do not happen while the bus is at the bus stop, they have to be recognised and filtered out to produce a better estimation of the arrival and departure times of the public transport vehicles.

***GPS errors***: The way the system recognises the stop where the transactions are occurring is to read the position of the buses from their GPS devices. If GPS readings are incorrect, especially when stops are very close to each other, during inclement weather or in high-rise urban environments, the stop identifier could be recorded incorrectly.

For a complete description of the trajectory reconstruction process, the reader is referred to Fourie (2014). These estimations of when dwell operations occur and the trajectories between stops were coded into MATSim 'events'; time-stamped, atomic units of information normally generated by the agent-based simulation that give a complete description of all vehicle and commuter agent actions during the course of the simulated day. The resulting XML file can be visualized and analysed using MATSim-compliant software and direct comparisons against MATSim simulations are greatly simplified.

## 4.2 Generation of a Public Transit Schedule

We used the reconstructed bus trajectories to determine the number of services and the time when the services start for every bus line in Singapore. As the vehicle identifier of each bus is known in the CEPAS data, we assigned the corresponding type of bus in the simulation, accounting for carrying capacity, doors operation mode, single or double decker configuration (which affects the bus dwell time). Fig. 2 summarizes this process. We compared these results with the commonly used Google Transit Feed Specification (GTFS) of the public transport system in Singapore. It recognised a significantly smaller number of services in CEPAS data: 4 bus lines were not found, 33 bus routes (different sequences of stops within a bus line) were not found and of the 91115 services specified in GTFS our reconstruction process recognised only 78515 services (86%). It is possible that a whole service is not visible due to lack of transactions or GPS errors as mentioned before. The difference in this comparison is still considerable, so the GTFS numbers could be overestimated. As reconstruction of the train trajectories presents even greater challenges than those for bus, because the transactions are recorded at the station entrance and not when the passengers enter and exit the vehicles, a similar reconstruction process has not been implemented at this point, but we intend to implement the method developed by Sun et al. (2012) in a future iteration. Consequently, the number of train services and their start times are directly obtained from the GTFS.

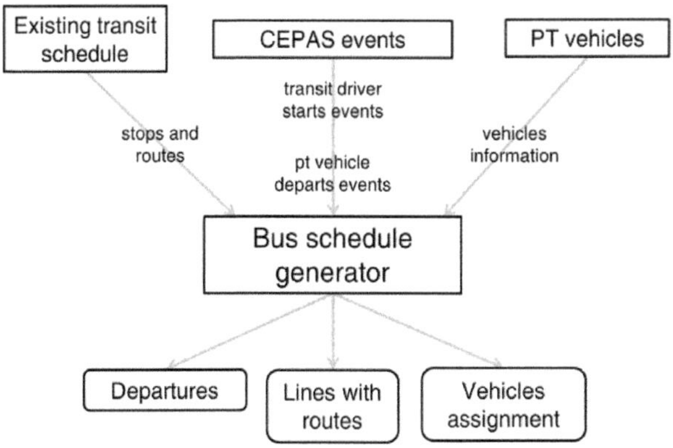


**Fig. 2.** Transit schedule generation using bus trajectory information from reconstruction process

## 4.3 Generation of Public Transport Trips

MATSim is an activity-based simulation framework and its demand description is a timed sequence of activity locations and connecting trips for each agent in the study area. Generating an agent-based demand description from the smart card data is a straightforward task; each boarding and alighting location could be used as an activity location in an agent's activity schedule. However, this would mean that we identify each transfer in a public transport trip as a significant activity and also over-specify the demand description by determining transfer location. It is important that realistic transfer locations, and their associated walking and waiting times, rather emerge from the simulation than be specified in the demand description. This end-to-end demand description requires identification of the initial boarding and final alighting location of each multistage trip in the smart card data and to use these transactions as approximate activity locations and activity start/end times of the agents. A number of challenges have been encountered in this process.

Access waiting forms an important component in an individual's transit experience; however, in the case of buses, recorded times don't correspond to user arrivals and departures to the public transport system. As transactions correspond to boarding and alighting only, the time when users arrive at the bus stop are unknown (except in transfers). More realistic bus stop arriving times for passengers are important for waiting time calculations. On the other hand, bus-bus, bus-train and train-bus transfer times are known and even exact bus lines could be assigned. That means, bus routes are fully reproducible from reality.

To assign bus users trip start times and identify individual multistage bus trips, we developed a two-step procedure. First, when a user alights from a bus and enters another vehicle, we established a threshold of 25 minutes to categorize those transactions as transfers or not transfers. If the time between alighting and boarding is more than 25 minutes, it is assumed that the user has left the system, therefore, they accumulate newly recorded access waiting time upon re-entry. The second step assigns bus users start times, using the reconstructed bus trajectories to extract headway times between consecutive services of the specified line. It has to assume (i) users wait exclusively for services of the line that they boarded in the transaction ignoring other lines that serve the same stops (ii) they do not have external information on bus arrivals. This is not always true as users could be waiting for more than one bus number. They also can have more information about reliable bus arrivals from experience, or digital apps which estimate bus arrivals. Given these assumptions we assigned a uniformly distributed user arrival time to the bus stop within the corresponding headway.

Thus, we generated a MATSim activity plan for each CEPAS user, assigning dummy activities between given or estimated arrival and departure times to the public transport system.

## 4.4 Simplification of the Network and Mobility Simulation

As only public transport vehicles are simulated, a detailed topology of the road network is not necessary. A reduction in the number of links and nodes of the road network represents a direct reduction in the MATSim mobility simulation computation time as its complexity is proportional to the network size and the number of agents. Thus, as Fig. 3 shows, a single link precedes each public transport stop (dwelling link) and a single link connects each pair of consecutive stops. If two stops are consecutive in at least one line then a link was created between them.

As mentioned before, the original MATSim mobility simulation is based on queues of vehicles at every link of the road network, depending on its corresponding capacity. That's how it accounts for the effect of car congestion on buses or vice versa. Without information about cars, but many observations of buses travelling, we introduced a stochastic travel time model, where the speed of buses on each link is drawn from a normal distribution; the parameters of which vary by time of the day and are the result of a multinomial regression model estimated from the speeds observed between stops from the trajectory reconstruction step (Fourie 2014). The parameters and the results of the regression estimation will be discussed in the following section.

With the modelled dynamic distributions, we modified the standard MATSim link dynamics (the queue model). Now, when a vehicle enters a link after a dwelling operation, a speed value of the link's distribution for the corresponding time of the day is sampled. During that time the vehicle "goes to sleep" and afterwards it appears at at the entry to the dwell operations link where it will queue up to allow passengers to board and alight. As our procedure does not reconstruct train trajectories yet, the simulation uses the standard queue model for the rail mode (as trains in subway systems have less interaction with other vehicles this approach is not far from the real behaviour).

## 4.5 Speed Regression Model

As shown by Sarlas and Axhausen (2015), the speed of vehicles in a network link are not only related to the level of demand on the link, but also to the network topology, presence of signalling systems and surrounding urban density and activity level. While their study calculated average travel times for the entire Swiss road network of private and public transport, our investigation focuses on determining observed speeds at any given time of day as function of network topology and indicators of the level of activity and demand that can be derived from the smart card data. The estimation results are shown in Table 1.

The model predicts the *natural logarithm* of speed (m/s) as a function of 15 variables listed in the table. Variable names in bold denote dynamic quantities that change on a per second basis. Derive all other variables from the network topology. The table shows the estimated value of the parameter, followed by the t-value. The last column shows the relative importance of the variable in terms of its contribution to the multiple R squared value listed at the bottom, using the method of Lindeman et al. (1980), implemented in the R statistical analysis platform (R Core Team 2014) by Grömping (2006).

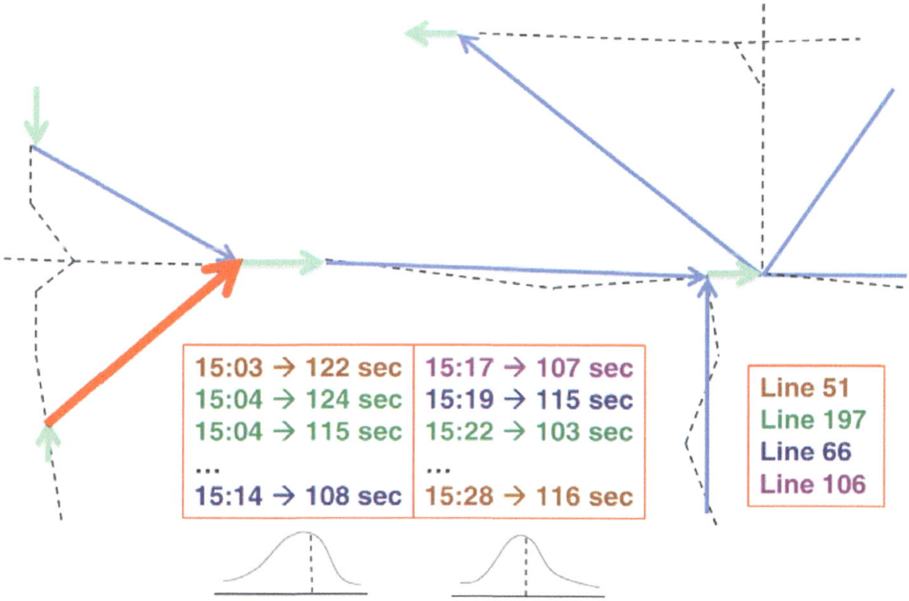


**Fig. 3.** Simplification of the MATSim network topology, showing stop to stop links and dwelling links before the stop

**Table 1.** Coefficient estimates of a multinomial regression model predicting the natural logarithm of bus speeds between stops (m/s)

| | Estimate | t-value | Relative importance |
|---|---|---|---|
| Intercept | 3.07E-01 | 6.48 | |
| Intersections per km | 5.07E-03 | 7.99 | 6.87% |
| Fraction of path with bus lane | 3.54E-02 | 9.17 | 0.49% |
| **Number of passengers tapped in** | **−1.55E-06** | **−20.13** | **3.38%** |
| Avg. number of intersections per roving sq. km | −7.32E-04 | −24.98 | 13.60% |
| Avg. degree of intersection nodes along path | −5.07E-02 | −14.77 | 4.67% |
| Right turns made at intersections | −7.46E-02 | −12.06 | 2.47% |
| Left turns made at intersections | −1.29E-02 | −2.06 | 0.28% |
| Right turns passed at intersections | −3.47E-02 | −14.47 | 2.67% |
| Left turns passed at intersections | −2.45E-02 | −10.56 | 1.59% |
| Number of nodes within traffic control buffer | −3.12E-02 | −30.25 | 8.99% |
| Path length (log) | 4.81E-01 | 64.87 | 21.81% |
| Number of arrivals at destination stop per day (log) | −3.72E-02 | −14.90 | 2.61% |
| Number of nodes in path (log) | −6.42E-02 | −11.71 | 2.97% |
| Path length over Euclidean distance (log) | −3.83E-01 | −24.19 | 4.33% |
| RMS radians turned (log) | −9.66E-03 | −4.78 | 2.19% |
| **Activities in progress per roving sq. km (log)** | **−2.28E-02** | **−11.26** | **6.19%** |
| **Smart card transaction rate per roving sq. km (log)** | **−7.28E-02** | **−29.90** | **14.90%** |
| **Multiple R-squared: 0.2054, Adjusted R-squared: 0.2053** | | | |

A Java class was created to calculate the topological variables, as well as to associate smart card transactions with network locations and use them to derive indicators of the level of activity and traffic that a bus might encounter between two stops at any given time of day. The variable names listed are relatively self-explanatory; often the natural logarithm of variables was used instead of their original values, in order for these variables to appear more normally distributed. Less self-explanatory variables are defined as follows:

***Intersections per km***: the number of nodes along the path of the bus between two stops that have more than one ingoing and one outgoing link or two pairs of parallel ingoing/outgoing links (nodes denoting changes in direction for one-way or bidirectional roads, respectively, therefore not intersections), divided by the length of the path in kilometres.

***Fraction of path with bus lane***: a number of road segments in Singapore have bus-only lanes in the leftmost lane, that are either exclusively for bus during the entire day or during peak hours. This variable denotes the fraction of the path that has such a bus lane. It does not take account of exclusivity by time of day, so this is a static variable.

***Total number of passengers tapped in system-wide***: This variable denotes the general accumulation of passengers in the entire public transport system and is therefore an indication of overall system load by time of day.

***Average number of intersections per roving sq. km***: For each node in the path between two stops, draw a circle with a 1 $km^2$ area and count the number of intersections within that area according to the definition stated earlier and then divide the sum by the number of nodes in the path.

***Average degree of intersection nodes along path***: For each intersection along the path count the number of links that meet in the node as a sign of its relative complexity; the more links that meet at an intersection affects the signaling times.

***Right turns made at intersections, etc.***: When a bus has to take a right turn at an intersection it generally takes longer than taking a left turn, as the estimates of these variables clearly reflect. In fact, making a left turn does not seem to have much effect on the model as reflected by its low t-value, despite the large sample size of more than a hundred thousand stop-stop combinations used in the estimation.

***Number of nodes within traffic control buffer***: The locations of traffic control signals were supplied by the Land Transport Authority as a geographically encoded shape file. This variable records the number of nodes in the path of the bus that fall within a buffer of 30 meters from a traffic control signal.

***Number of bus arrivals at destination stop per day*** (*log*): This variable accounts for the traffic at the destination stop, with the expectation that the more services offered at the stop, the longer a bus is likely to wait in a queue before it can perform dwell operations.

***Number of nodes in path, path length over Euclidean distance, RMS radians turned***: These variables attempt to capture the degree of 'friction' between the two consecutive stops that prevent the bus from reaching top speed.

***Activities in progress per roving sq.km*** (*log*): For each node in the path between the two stops, draw a circle with a 1 $km^2$ area around the node and retrieve all smart card transactions recorded at the public transport stops within the circle. It assigns each boarding transaction a value of -1 and each alighting transaction a value of +1 and finds the running sum of the values by time of day. We subtract the minimum of the running sum from all its values and uses the resulting set of values as an indication of the number of activities that take place within the circle. For a given time of the day the value of the running sum at each node is read from a table and the average of these values across all nodes in the path is used in the regression.

***Smart card transaction rate per roving sq. km (log)***: For each node in the path between two stops at a given time of day, we calculate the 15 minutes moving average of the total number of transactions taking place per second within a 1 $km^2$ circle around the node and use the average of these values across all nodes in the path. This value is used as a sign of the general level of traffic that the bus encounters along its path between stops.

A correlation analysis of the variables used in the estimation shows that some variables have high degrees of correlation for obvious reasons; for instance, the number of left and right turns taken at intersections are obviously dependent on the number of intersections encountered. The variables describing the degree of 'friction' encountered along the path are also positively correlated, while path length and number of nodes within traffic control generally increase with the number of nodes in the path. Exclusive bus lanes also appear to be associated with stops with a large number of services arriving per day.

The model might therefore suffer from a significant degree of multicollinearity; however, estimated variable coefficients are stable for different sample sizes and adding new variables to the model do not lead to erratic changes in estimated values. But correlated variables could, arguably, be replaced by their first principal components to remove collinearity effects while retaining predictive power, as done in principal component regression.

The current estimation of the model does not take account of spatial autocorrelation. An initial investigation of the residual (using a k-nearest neighbour approach on the destination stop to define the neighbourhood matrix used in spatial autocorrelation models) does show a significant degree of spatial autocorrelation that varies by time of day, with a maximum value of Moran's I of 0.12. Effects of spatial autocorrelation will be further investigated in future studies, however as would be seen in the validation section to follow, the current ordinary least squares estimation already gives very reasonable predictions of speed and resulting passenger travel times.

The MATSim link dynamics model uses the predicted logarithm value of speed from the regression model for the given time of the day as the mean for the normal distribution to sample the final speed value from and the standard deviation of the distribution used for sampling is that of the residual for predicted speeds within 0.5 km/h of the mean.

## 4.6 Dwell Time Model

In Sun et al. (2013) the authors show how different bus configurations translate into different rates of boarding and alighting. Furthermore, from a study of the Singaporean smart card data and knowledge of the bus type associated with each vehicle identifier in the data, they derived a model of

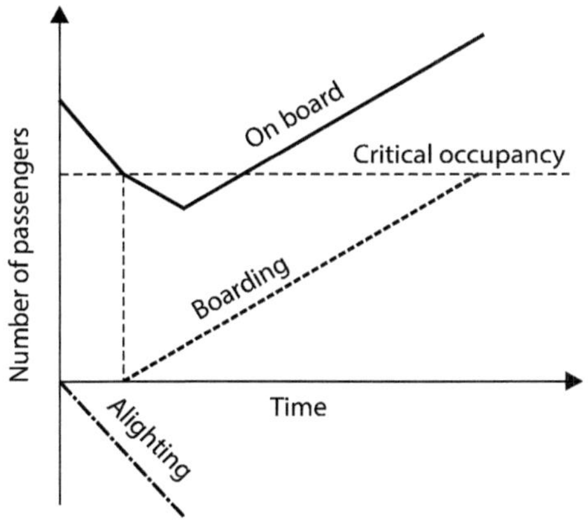


**Fig. 4.** Dynamics of passengers boarding and alighting from a crowded bus, revealing how boarding can only precede once a critical occupancy has been reached

dwell time variability as a result of boarding and alighting transactions and the number of passengers on board the bus. The most significant effects could be observed when the bus is full and it is impossible for passengers to board until enough passengers have alighted, as could be seen in Fig. 4.

Hence, this variable dwell-time model was incorporated in the MATSim simulation. As will be seen later in the validation section, simulated results from the model fits very well with observed values.

## 5. VALIDATION AND PERFORMANCE

We ran the modified MATSim simulation model for 50 iterations and compared various measures of system performance against the original smart card data and the trajectory reconstruction-related data.

### 5.1 Speed

Fig. 5 compares distribution of the bus speeds between stops in the simulation against the speeds derived from the trajectory reconstruction process. Both the shape of the distributions and absolute numbers correspond very well.

### 5.2 Headways, Dwell Times and Bus Bunching

Figure 6 shows the distribution of the headways in the simulation versus those derived from the trajectory reconstruction process. In its current state the simulation appears to produce too many short headways; this is due to somewhat excessive bus bunching that occurs during the simulation, reducing the headway between consecutive buses to nearly zero.

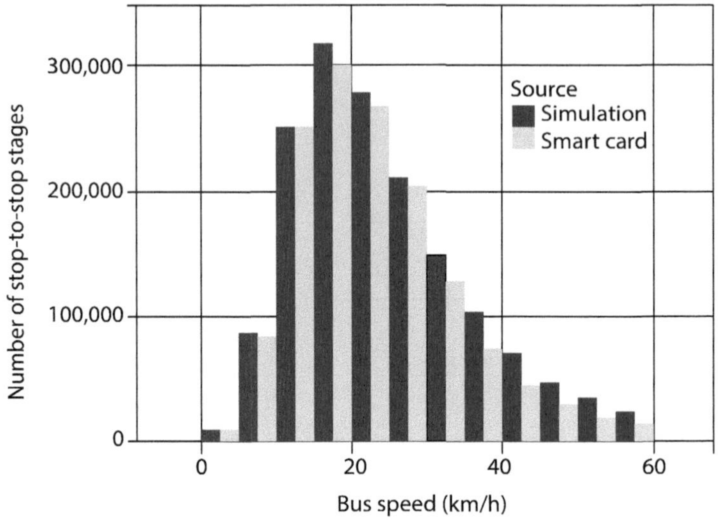


**Fig. 5.** Distribution of bus speeds from the smart card trajectory reconstruction process and the simulation

In terms of the headway variability, it shows that the simulation produces increasing headway variability with increasing number of stops along the route, however the effect is much more pronounced in the simulation.

It could be seen from the joint distribution of headway versus number of stops along the route that the simulation produces many headways in the 0-1 minute bin, which indicates bus bunching. This behaviour in the simulation is probably largely due to the first-in-first out queue dynamics of the simulation that prevents buses of the same service from passing each other. Therefore, we intend to investigate passing behaviour in future iterations, as buses of the same service can pass each other in reality when the first bus is already engaged in a dwell operation at a stop.

As the trajectory reconstruction process does not extrapolate the trajectories beyond the last recorded transaction for a circuit run, headways for the stops towards the end of a route might be inaccurate, which accounts for the lighter shading of the joint distribution of headway versus stop number in the smart card data. However, the distribution of the headways does appear considerably narrower for the smart card data than what the simulation produces. We are not aware of any bus bunching control measures in operating the buses, whether it is centralized control from the operation centre, or by intelligent actions of the bus drivers themselves. Such measures would naturally account for the increased reliability of services. However, it would also be worth investigating if allowing buses of the same service to pass each other when one bus is already occupied at a bus stop, serves as a bunching control measure in itself.

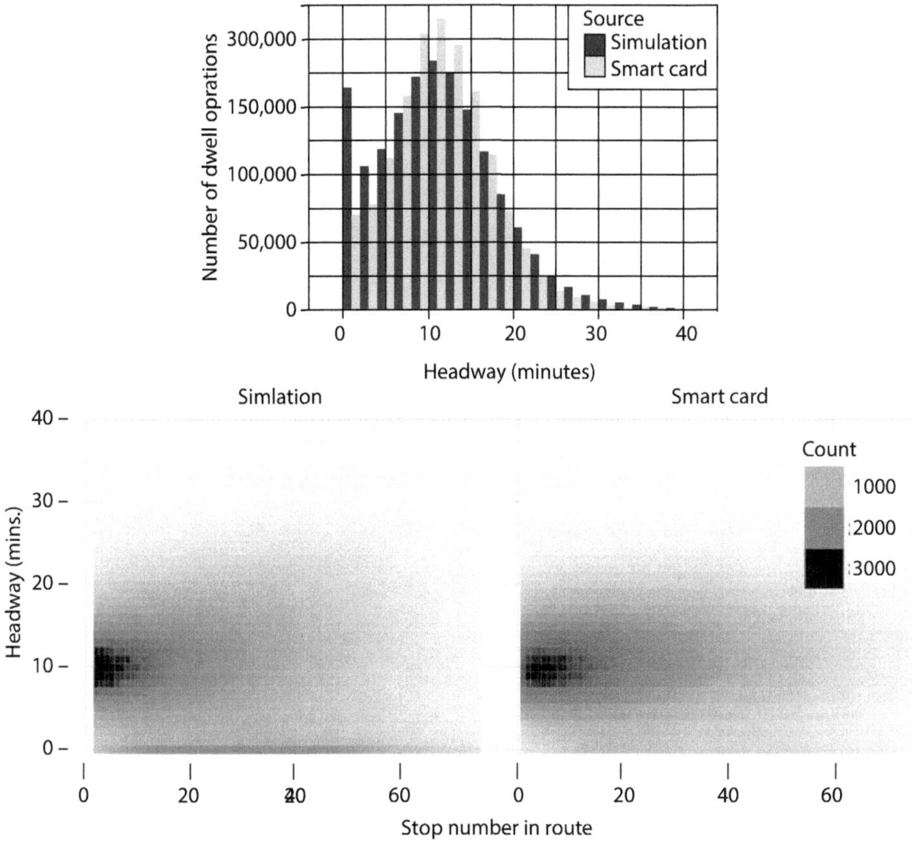


**Fig. 6.** Comparison of the bus headway distributions and headway variability with increasing stop numbers along the route

In Figure 7 the dwell time of buses in the simulation is compared against those derived from the trajectory reconstruction process. In terms of absolute numbers, nearly 1 million dwell operations with zero length occur in the simulation; these are cases where no boarding or alighting transactions take place. In the trajectory reconstruction, dwell operations that have been interpolated are assigned a zero dwell time. Dwell operations where only a very small number of transactions were recorded within a time span of less than six seconds, were assigned an arbitrary lesser dwell time of that value, which is responsible for the second spike in dwell times that could be seen in the histogram. In terms of absolute numbers, the sum of these trivial cases for the smart card data corresponds reasonably well with the number of dwell operations in the simulation where no passengers board or alight.

Because the absolute numbers of dwell operations for the non-trivial cases are different for the simulation and the smart card data, we compared the distributions in terms of their density in the second part of the figure,

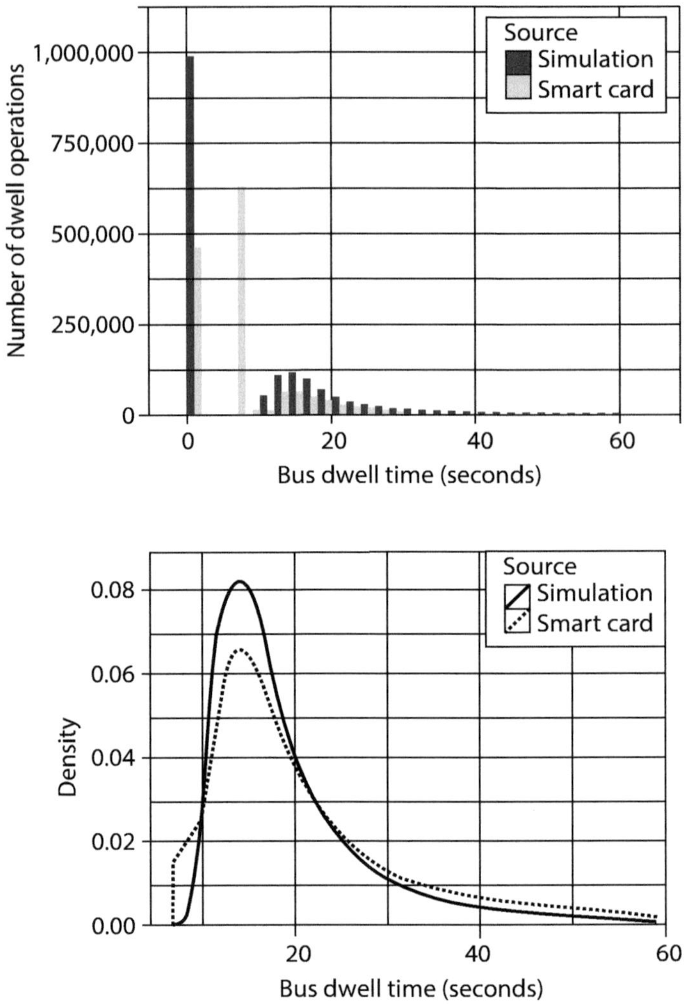


**Fig. 7.** Bus dwell time histogram and density comparison of non-trivial cases

which reveals reasonably good correspondence in terms of distribution between the simulation and the dwell times from the trajectory reconstruction process.

The trajectory reconstruction process produces 1.58 million dwell operations compared with the 1.7 million dwell operations recorded in the simulation; the difference between these numbers is due to the trajectory reconstruction process not extrapolating the trajectories of buses beyond the first and last recorded transactions. So, if the first recorded transaction for a bus occurs at a stop after the first in its route profile, or the last recorded transaction is before the end of the line, then no dwell operations were created for the stops before the first transaction, or after the last transaction.

Figure 7 also shows that, of the 1.7 million dwell operations in the simulation, nearly one million have zero duration, meaning that no passengers were picked up or dropped off. This means that buses in the simulation only pick up or drop off passengers approximately 40% of the time. Consequently, the simulation also produces more dwell operations of longer duration, as fewer dwell operations have to serve the same number of passengers. This might be a contributing factor to the higher incidence of bus bunching observed in the simulation.

If we assume that the actual total number of dwell operations also comes to 1.7 million, then the number of cases where buses don't take on any passengers at stops for that particular day in the actual transport system which comes to approximately 580,000, which accounts for approximately 34% of all dwell operations, meaning that buses in reality pick up or drop off passengers 66% of the time, in comparison to the 40% observed in the simulation. This difference might be due to the best response routing in the simulation resulting in increased coordination between the agents and the buses, with agents selecting services that get them to their destination with less access waiting time on average than the service that they picked in reality. Agents might also not be as averse to crowding as people in reality, causing them to opt for the next empty vehicle less often; a hypothesis that will need further investigation into the ridership of vehicles in the simulation versus those in reality.

The space-time diagram shown in Fig. 8. compares the trajectory reconstruction results against the simulation for a bus line with 74 stops along its route. While the shapes of the trajectories compare reasonably well, it could be seen that the simulation produces more bus bunching than what this bus line has experienced in reality, confirming what is apparent in the histogram in Figure 6.

## 5.3 Passenger Travel Time Measures

Fig. 9 compares the simulation trip travel time with the smart card data, where access and egress walking and waiting times have been extracted from the times recorded in the simulation. The histogram therefore compares only the sum of in-vehicle travel times and transfer walking and waiting times.

Fig. 10 similarly shows very good agreement between the bus stage in-vehicle times for the simulated versus real values, although smart card values appear slightly skewed to longer times. While the simulated speeds are stochastic, to display the same range of values as those observed in reality, it is possible that not all dynamic effects have been adequately captured captured for perfect agreement, or the agents are routed more optimally than passengers are in reality. As we do not know when passengers board or alight from trains, we cannot construct a similar graph for rail modes. However, the good agreement observable for trip travel time across all modes gives confidence that simulation of the rail

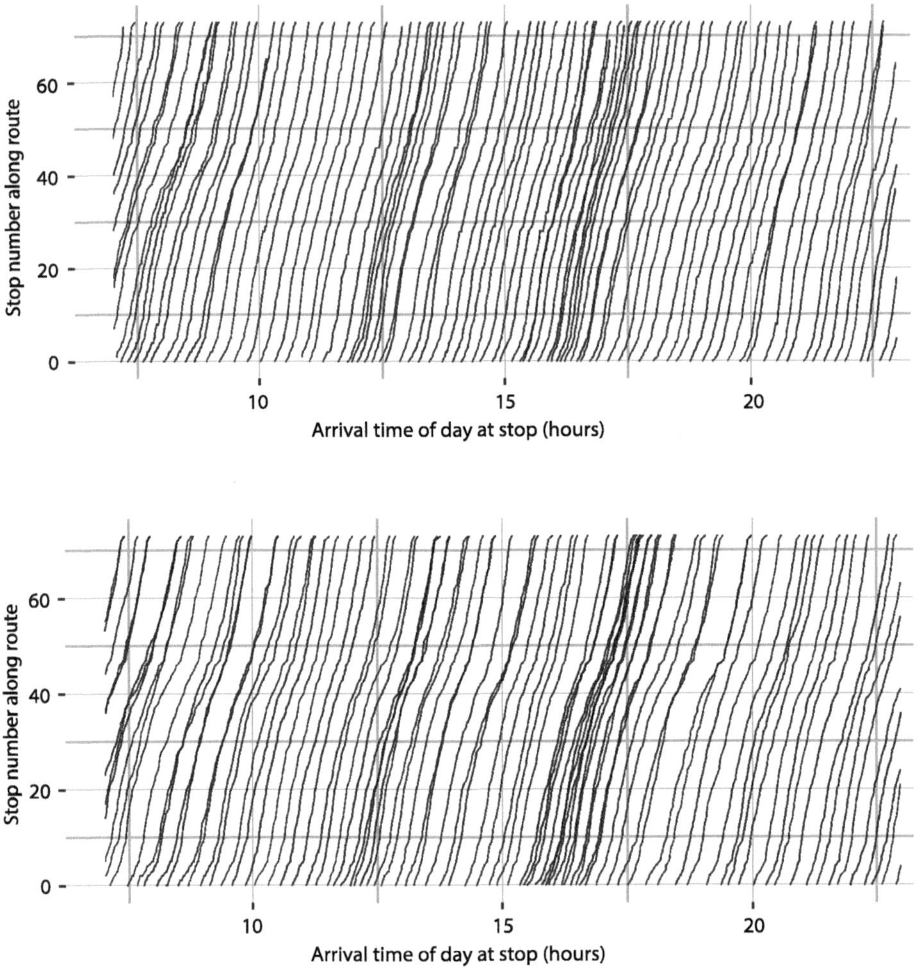


**Fig. 8.** Comparison of space-time trajectories of CEPAS (top) versus simulation (bottom)

mode is reasonably accurate, as passengers would have switched away or switched to using the subway during the simulation if this transport mode performed markedly different from reality.

Fig. 11 compares density of the transfer times in the simulation against those derived from the smart card data. In this case we do not display the histogram of transfer times, as the absolute numbers inferred from the smart card data are inaccurate; especially for the train mode we do not know exactly which routes passengers have taken, or exactly how long they have spent in transfer. The absolute numbers suggest that times in MATSim might be somewhat shorter than those experienced in reality, possibly due to the co-ordination that occurs due to best response re-routing during the simulation, as alluded to earlier.

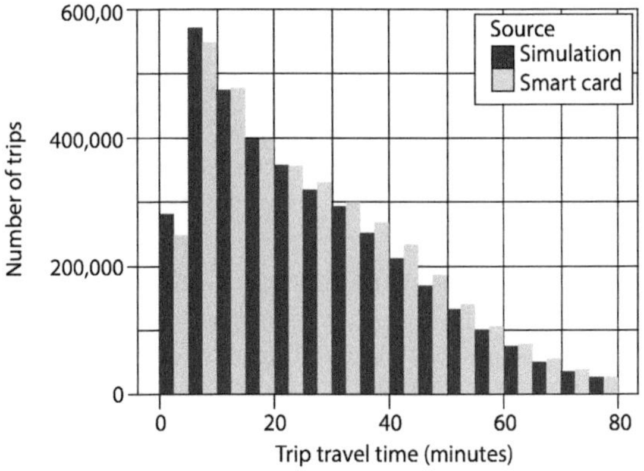


**Fig. 9.** Comparison of simulated versus real trip travel times across all modes of public transport (excluding access and egress times)

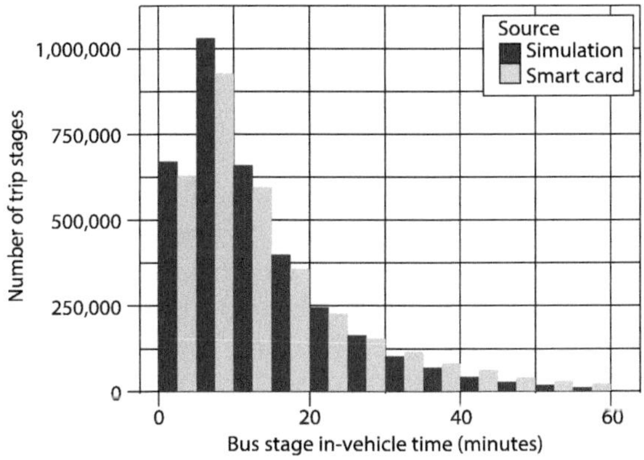


**Fig. 10.** Comparison of simulated versus real bus stage-in vehicle time

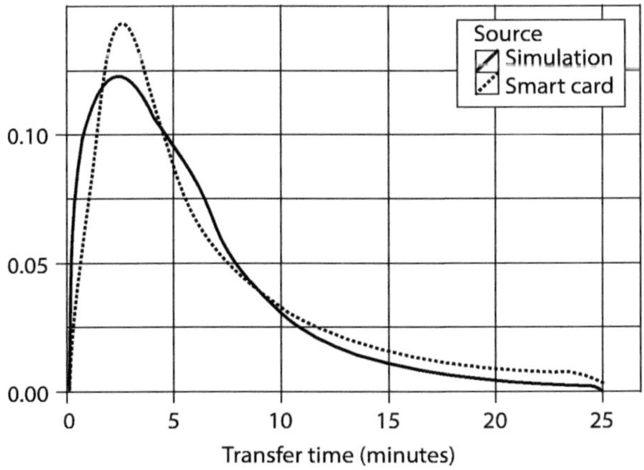


**Fig. 11.** Comparison of transfer time density in the simulation against that derived from the smart card data

## 5.4 Computation Times of Simplified Simulation

Using only best response re-routing, the simulation reaches a relaxed state in very few iterations. After only five iterations very little change in the average score of agent plans could be observed with increasing iterations. From experience we found that one only needs to run a 25% sample of all agents to get realistic results; all counts recorded in the validation section were thus from such a sample and were scaled up by multiplying them by four.

The experiments were run on a latest generation 24 core Intel Xeon computer, with 64 GB of RAM. The initial routing of all agent plans takes approximately seven minutes, while a single iteration takes approximately four minutes. It is therefore possible to have usable results in under an hour. In the case of a standard MATSim simulation of both public and private transport, many more iterations are required for the system to reach a relaxed state, and a full simulation can take up to two days to complete. The simplified simulation therefore represents a big step forward in terms of computation time performance.

# 6. APPLICATION

To show the potential of the simplified public transport simulation we designed a fictitious case study. In the proposed scenario we split one of the longest bus lines in Singapore, which has more than 90 stops. The line was split according to the method used in Lee et al. (2012) in order to minimize the number of transfers resulting from the split; in this case the optimal split point happened to be close to the centre of the route. Agents were allowed to re-route their public transport routes within the MATSim co-evolutionary algorithm until they reach equilibrium (100 iterations). That means the agents who were taking the long line or any other line in Singapore can decide to take the new split line or switch to another transit line. As in the case of the validation study, we simulated a 25% sample of the population, with vehicle carrying capacities reduced to a quarter of their real-world values. The following section compares performance of the line split against the baseline case.

## 6.1 Impact on Bus Bunching

Fig. 12 shows the space-time diagram of the bus service before and after the split, with cases of bus-bunching highlighted in red and line thickness increasing with bus ridership. The plot confirms that the incidence of bus bunching is significantly reduced during the morning peak hour and that headway reliability improved considerably, especially towards the end of the bus route. Note that we replicated departure times from the start of the service for buses departing on the second part of the line split, which means that these services start with an inherent lack of reliability.

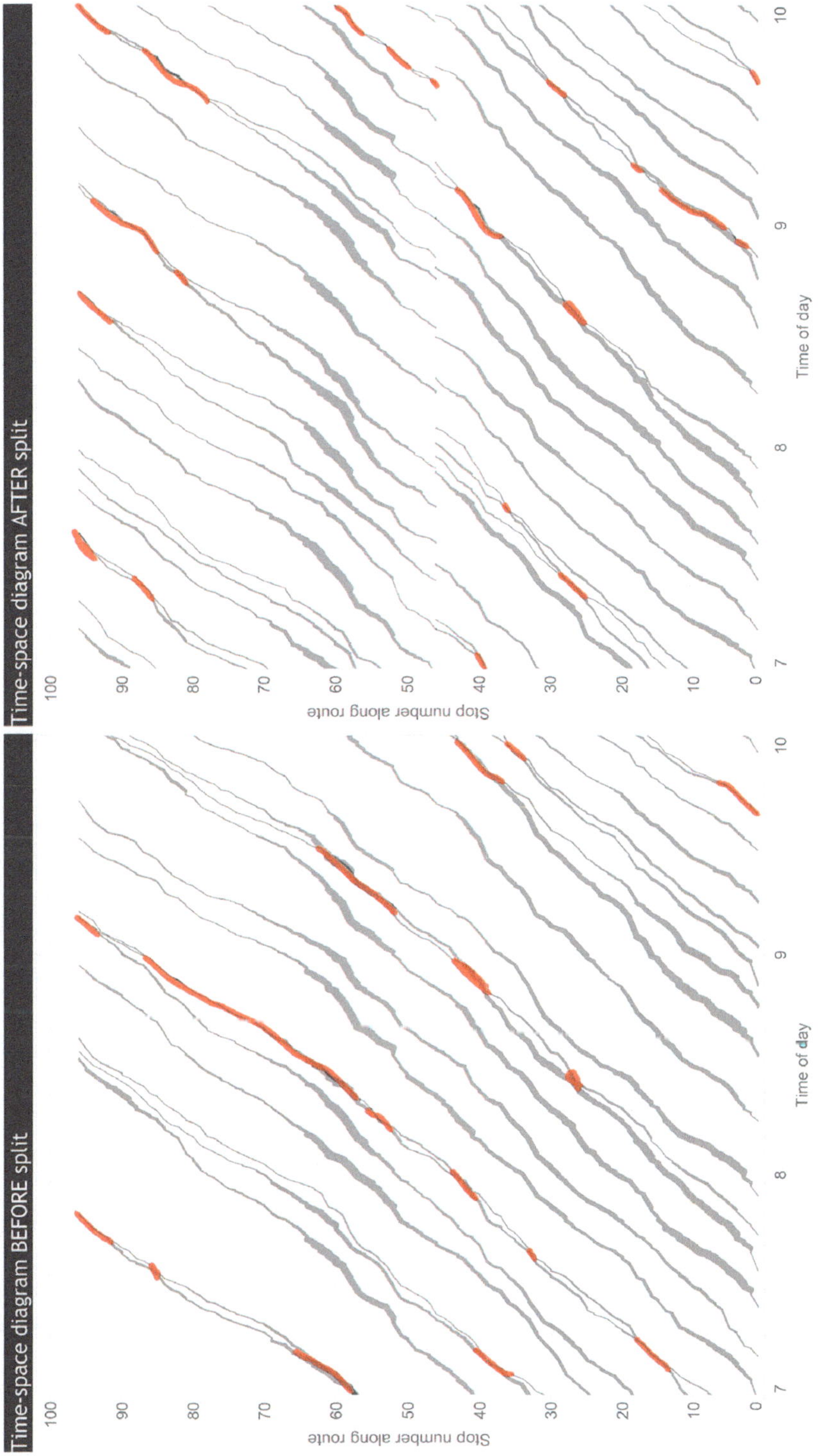


**Fig. 12.** Space-time diagram of a bus line before and after the line has been split in two

Furthermore, even though we reduced the number of stops in the two resulting routes, it is clear in the base case that bus bunching can result relatively early and that 45+ stops might still be too many bus stops for a reliable bus service.

## 6.2 Excess Waiting Times

Excess waiting time (EWT) is one of the most common reliability indicators for high frequency public transport services (e.g., a service frequency of five or more buses per hour). Using the definitions used by the London transport authorities, EWT assessment includes calculation of the following two elements:

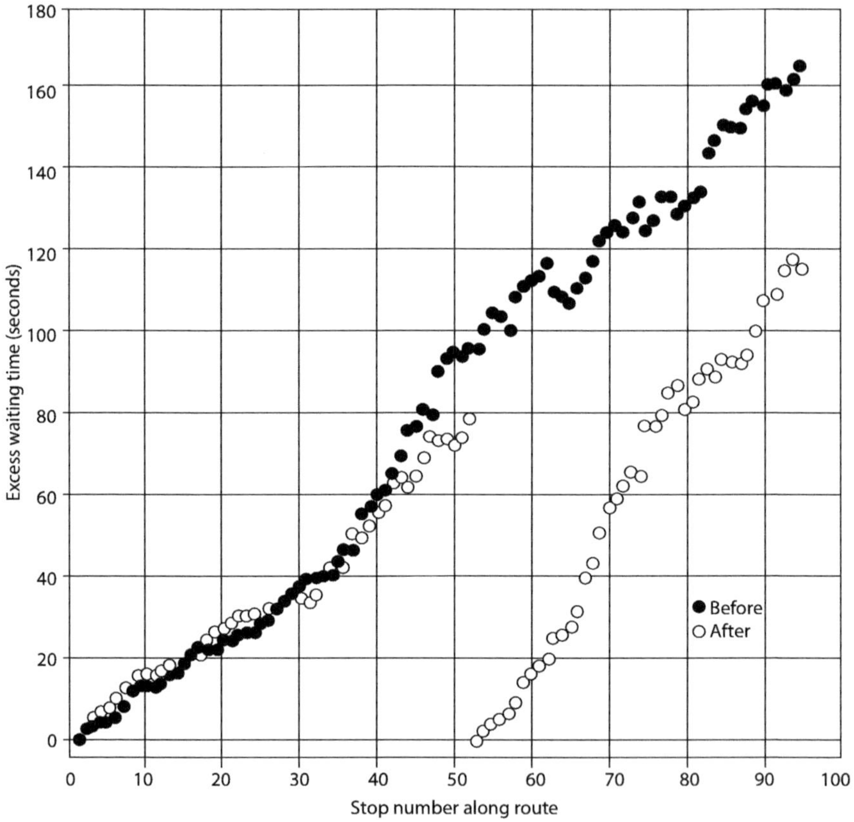


**Fig. 13.** Comparison of the excess waiting time before and after a long bus line has been split into two separate routes

***Average scheduled waiting time (SWT)***: the time passengers would wait, on an average, if the service ran exactly as scheduled, assuming that waiting time is, on average, half of headway time:

$$SWT = \frac{\sum_{s \in S} H_s^2}{2\sum_{s \in S} H_s}$$

***Average actual waiting time*** (*AWT*): the average time that passenger actually waited,

$$AWT = \frac{\sum_{s \in S} H_a^2}{2\sum_{s \in S} H_a}$$

where $s$ represents each service of a bus line (excluding the first one), $H_s$ is the scheduled headway of the service $s$ and the previous service, and $H_a H_s$ is the real headway of the service $s$ and the previous service. EWT is simply the difference between AWT and SWT and represents the additional waiting time experienced by passengers.

The formulas have this form because AWT and SWT are weighted averages of all the service headways of a line and the weight is the real headway. So, if the line is designed to have a constant headway, the calculation of SWT could be simplified to $SWT = 0.5H_s$.

Fig. 13. compares the calculation of the EWT of the base case against the split line scenario, in one direction of travel. The plot looks very similar in the opposite direction; EWT reverts to zero at the point with the line split so passengers experience better reliability towards the end of the route.

## 7. CONCLUSION

From the section on validation, the results so far seem to agree well for most part with observation. Most importantly, the simplified simulation manages to capture dynamic bus bunching effects; in fact, the effect might be slightly exaggerated in the simulation. The possibility of mitigating this effect through the implementation of passing behaviour in the queue simulation should be investigated. The simple fictitious case study also illustrates that the simplified simulation could be used to evaluate proposed changes to the public transport system.

The reconstruction of train trajectories is a very interesting problem as train-to-train transfers are not explicit in the CEPAS data. Furthermore, public transport passengers need to be located to buildings that are close to public transport stops, to better simulate access walking and waiting times.

The subway stops are also easily accessible in the simulation and do not take account of the time that it takes for passengers to travel all the way down to station entrances. Consequently, there is a slightly increased preference for the rail modes in the simulation compared to reality.

It was our experience that the first-order analyses and operations on the smart card data, such as the conversion to trajectories, the speed regression and the models of boarding and alighting, are relatively straightforward to implement. The task of integrating results into a simulation model capable of providing insight into future transport system performance was a radically more involved task. The bugbears of systems engineering, namely unanticipated interactions and emergent phenomena, come into play even in this highly simplified integrated model, because of the dynamic and disaggregate interacting nature of the agent-based simulation. In design iterations leading to the current state of the system, it took many hours of tinkering with its components in order to isolate cause and effect and a number of challenges still remain, as highlighted throughout the section on validation of results.

Whether the integrated modelling approach ultimately proves worthwhile in predicting future transport system performance or not, the value of smart card data during all stages of the design and evaluation is undeniable. Whenever confronted with unexpected behaviour in the simulation, we found ourselves constantly turning to the data for answers, trying to infer what actually happens in reality. We expect that this will become even more the case as data could be potentially enriched with bus GPS traces and the mystery of where passengers are in the train system during the time between transactions at station entrances is revealed through rigorous statistical analyses and the promise of co-ordinated data from underground cell phone transceivers.

## ACKNOWLEDGEMENTS

This research is supported by the National Research Foundation of Singapore. Input data was supplied by the Land Transport Authority of Singapore, unless otherwise indicated.

## REFERENCES

Bagchi, M. and White, P. 2004. What role for smartcard data from bus systems? *Municipal Engineer*, 157, pp. 39-46.

Bagchi, M. and White, P.R. 2005. The potential of public transport smart card data. *Transport Policy*, 12, pp. 464-474.

Balmer, M., Rieser, M., Meister, K., Charypar, D., Lefebvre, N. and Nagel, K. 2009. MATSim-T: Architecture and simulation times. *Multi-Agent Systems for Traffic and Transportation Engineering*, pp. 57-78.

Charypar, D. and Nagel, K. 2005. Generating complete all-day activity plans with genetic algorithms. *Transportation*, 32, pp. 369-397. doi:10.1007/s11116-004-8287-y.

Fourie, P.J. 2014. Reconstructing bus vehicle trajectories from transit smart card data. Working paper 986, ETH Zurich, *Institute for Transport Planning and Systems*.

Grömping, U. 2006. Relative importance for linear regression in R: A vignette for relaimpo.

Lee, D.-H., Sun, L. and Erath, A. 2012. Determining Optimal Control Stop to Improve Bus Services Reliability. *Presented at the 1st European Symposium on Quantitative Methods in Transportation Systems*, Lausanne, Switzerland.

Lindeman, R.H., Merenda, P.F. and Gold, R.Z. 1980. Introduction to bivariate and multivariate analysis. Scott, Foresman Glenview, IL.

Munizaga, M.A. and Palma, C. 2012. Estimation of a disaggregate multimodal public transport Origin–Destination matrix from passive smart card data from Santiago, Chile. *Transportation Research Part C: Emerging Technologies*, 24, pp. 9-18. doi:10.1016/j.trc.2012.01.007.

Pelletier, M.-P., Trépanier, M. and Morency, C. 2011. Smart card data use in public transit: A literature review. *Transportation Research Part C: Emerging Technologies*, 19, pp. 557-568. doi:10.1016/j.trc.2010.12.003.

Prakasam, S. 2008. The Evolution of e-payments in Public Transport – Singapore's Experience. *Japan Railway and Transport Review*, 50, pp. 36-39.

R. Core Team. 2014. R: A language and environment for statistical computing. *R Foundation for Statistical Computing*, Vienna, Austria, 2012. ISBN 3-900051-07-0.

Rieser, M. 2010. Adding transit to an agent-based transportation simulation. *Ph.D. Thesis*, Technical University Berlin, Berlin.

Sarlas, G. and Axhausen, K.W. 2015. Localized speed prediction with the use of spatial simultaneous autoregressive models. *Presented at the 94th Annual Meeting of the Transportation Research Board.*

Smith, L., Beckman, R. and Baggerly, K. 1995. *TRANSIMS: Transportation analysis and simulation system*. Los Alamos National Lab., NM (United States).

Sun, L., Lee, D.-H., Erath, A. and Huang, X. 2012. Using Smart Card Data to Extract Passenger's Spatio-temporal Density and Train's Trajectory of MRT System, in: *Proceedings of the ACM SIGKDD International Workshop on Urban Computing*, UrbComp '12. ACM, New York, NY, USA, pp. 142-148. doi:10.1145/2346496.2346519

Sun, L., Tirachini, A., Axhausen, K.W., Erath, A. and Lee, D.-H. 2013. Models of Bus Boarding/Alighting Dynamics and Dwell Time Variability. *Transportation Research Part A: Policy and Practice*, 69: pp. 447-460.

## AUTHOR BIOGRAPHY

**Pieter Fourie** is a simulation modeller at the Future Cities Laboratory in Singapore. His interests are focused on improving the performance of agent-based transport simulation, and simplifying its implementation through the use of alternative data sources, such as transit smart card and mobile phone data. He has used bus smart card data from Singapore to reconstruct bus trajectories and developed models of bus speeds. This work has has been implemented in the first smart card driven agent-based simulation described in this chapter.

**Sergio A. Ordonez M.** is a Colombian computer scientist and mechanical engineer specialized in modelling and simulation at the Future Cities Laboratory in Singapore. In his dissertation he addresses the problem of simulating large scale urban transport scenarios during long periods of time, to study work-leisure human cycles. Multi-day public transport smart card records and common travel surveys are the main sources of his activity-based model.

**Artem Chakirov** is an associate researcher in the areas of Mobility and Transport Planning at FCL. His current work focuses on mobility pricing in urban areas. Previously Artem was also involved in demand generation for Singapore transportation model and analysis of public transport smart card data.

**Dr. Alexander Erath** leads the Engaging Mobility group at the Future Cities Laboratory (FCL) of the Singapore-ETH Centre. In this role, he led the first implementation and further development of the large-scale, agent-based transport demand model MATSim Singapore and initiated the idea of using Smart Card Data for agent-based simulation. His main research interests are surveying and modelling of travel behaviour such as quantifying the impact of the built environment on mobility and transport demand modelling.

**Dr. K.W. Axhausen** is Professor of Transport Planning at the Eidgenössische Technische Hochschule (ETH) Zürich. He has been involved in the measurement and modelling of travel behaviour for the past 30 years contributing especially to the literature on stated preferences, micro-simulation of travel behaviour, valuation of travel time and its components, parking behaviour, activity scheduling and travel diary data collection. Current work focuses on the agent-based micro-simulation toolkit MATSim (see www.matsim.org).

# PART 3

# Smart Card Data for Evaluation

Chapter 9

# Smart Card Data for Wider Transport System Evaluation

M. Munizaga[1,*], C. Núñez[1,2] and A. Gschwender[1,2]

**ABSTRACT**

This chapter discusses the use of smart card data for evaluation of the transport system from the user's perspective. Tools are presented that, using passive data from the public transport system, can help transport planners to achieve their goal of improving mobility and quality of life. Using as an example the case of Santiago, Chile, it is shown how detailed analyses can be performed using passive data, contributing to the understanding of the system at a very low cost. The existence of broad and precise temporal and spatial data, allows making analyses at different levels of aggregation that are critical to understand the quality of service of urban public transport systems.

## 1. INTRODUCTION

The evaluation and monitoring of transport systems is critical for planning and for supporting decision making related to a city and its transport system. Traditionally, these procedures have been supported with survey data that contain detailed information based on respondents' declarations and measurement data of specific aspects of the transport systems of interest. However, due to the high costs of surveys and measurements, these methods are typically sparse in terms of time and space coverage. Conversely, the incorporation of technological devices in the daily operation of a public transport system has provided significant

[1] Departmento de Ingeniería Civil, Universidad de Chile, Blanco Encalada 2002, Santiago, Chile. Email: mamuniza@ing.uchile.cl
[2] Directorio de Transporte Público Metropolitano, Moneda 975, Piso 4, Santiago, Chile.
* Corresponding author

quantities of passive data. These databases are provided (and limited) by the technological devices available, which generate large databases with regard to specific aspects of passenger and vehicle movement. Additionally, because these devices require limited human intervention, they are usually particularly reliable in what they detect; however, what the technological devices detect is not necessarily what planners and decision makers require. The objective of this book chapter is to show that it is possible to estimate the level of service, mobility and accessibility indicators using passive data. Specifically, from a real case in the public transport system of Santiago, Chile, we will focus on the use of passive data from automatic fare collection (AFC) and automatic vehicle location (AVL) systems that are complemented with geographic information system data (GIS).

Bagchi and White (2005) recognise the potential of smartcard data to analyse boarding transactions, time-space distribution and turnover rates over time; however, they have also identified a series of problems. In AFC systems where passengers are not required to tap their cards when exiting the transport system, trip destination information is not available. In addition, no journey purpose, attitudes or quality of service information is recorded, and trip identification is based on rules that require validation. Researchers have observed these limitations as opportunities, and different authors have proposed methods to estimate an alighting bus stop (Zhao et al. 2007; Trepanier et al. 2007; Munizaga and Palma 2012), to link trips and analyse transfer behaviours (Seaborn et al. 2009; Charikov and Erath 2011; Devillaine et al. 2013) and to estimate a trip's purpose (Charikov and Erath 2011; Devillaine et al. 2013; Lee and Hickman 2014). Certain authors have also analysed travel behaviours at different levels: walking access behaviour (Utsonomiya et al. 2006), travel patterns and variability (Morency et al. 2007; Ma et al. 2013) and the location of regular activities (Charikov and Erath 2011; Amaya and Munizaga 2014).

In the next section, the levels of service indicators found in the literature are briefly discussed. Then, using passive data, the construction of the different levels of service indicators for the public transport system of Santiago is explained and shown at different aggregation levels. The final section presents some concluding remarks.

## 2. LEVEL OF SERVICE INDICATORS

Measuring a public transportation system's level of service has always been a concern for planners, operators and regulators. Initially, indicators focused on effectiveness and efficiency (Fielding et al. 1985). Based on TRB (2003), public transport indicators can be divided into performance or operation indicators and level-of-service indicators. Eboli and Mazzula (2012) discussed the fact that level-of-service indicators may vary depending on whether the operator, user or community perspective is taken. In this chapter, we focus on the perspectives of the user and of the potential users to evaluate the attractiveness of the public transport system and the stability of demand.

Considering the attributes that are relevant to measure the quality of service from the user's perspective, the first and most significant source of information is the list of attributes that affect user satisfaction (or perception). The literature related to factors influencing user satisfaction is extensive (Ortúzar et al. 1997; dell'Olio et al. 2010; Hensher et al. 2003; Tyrinopoulos and Antoniou 2008; Yañez et al. 2010 and Donoso et al. 2013). Another source of information regarding factors that affect the level of service from the user perspective is provided by the factors that determine users' route choices within the public transport network (e.g., dell'Ollio et al. 2011, Raveau et al. 2014 and Chapter 4 this book). Donoso et al. (2013) argue that the relative valuation of the factors that influence route choice may differ from the ones obtained when measuring user satisfaction. This difference is because the decisions (i.e., route choice or survey response) are usually made under different constraints, even though in both cases, the underlying function is the same (i.e., latent individual utility function); this may affect the estimation of the marginal utility of a specific factor. Despite this fact, the attributes that systematically appear to be relevant are:

- Travel time including walking, waiting, in-vehicle, and transfer
- Frequency compliance
- Coverage
- Seat availability
- Regularity of waiting and travel time
- Number of stages/transfers
- Public transport travel time compared to car travel time

The literature provides examples of the calculation of level-of-service indicators from passive data. Among the most relevant, Bertini and El-Geneidy (2003) calculated frequency, regularity and accessibility indices and load profiles for the public transport system of Portland (Oregon, USA). Similar calculations have been performed by Trépanier et al. (2009) for Gatineau (Canada). Bagchi and White (2005) calculated trip rates, transfers within linked trips, and turnover rates for the public transport systems of two cities in the UK. Additionally, Utsunomiya et al. (2006) calculated access distances and examine the travel patterns of users. In systems where tap-off or exit validation is also required, the scope of indicators that can be calculated increases. Park et al. (2008) calculated indicators including transfer rates and travel times for the public transport system of Seoul, Korea, which were complemented by Jang (2010) with the identification of critical transfer points.

## 3. APPLICATION TO SANTIAGO

The Santiago public transport system, which is called Transantiago, is a multimodal integrated system (i.e., bus and metro) that serves a population of 6.6 million inhabitants. It has over 6,000 buses, all equipped with GPS devices, operating daily in a network that contains 70 km of segregated bus ways and over 11,000 bus stops. The integrated metro network has 5 lines, 103 km of rails and 108 stations, and is currently expanding. The fare scheme is based on the trips taken by users; a flat fare is applied to trips through a maximum of three stages and must be used within two hours. A small surcharge, which is higher during peak-use hours, is applied to trips that use the metro network. The payment system is based on a contactless card called "bip!", which is the only method to pay in the system's buses and by far the most popular method to pay in the metro, representing 97% of the payment transactions of the 4.6 million daily trips via public transport. Given this fare structure, tap-off validation is not required in buses or the metro.

Previous work with these data has generated origin-destination matrices at the bus stop level that are based on the alighting estimation method proposed by Munizaga and Palma (2012) and the trip-stages linking procedure proposed by Devillaine et al. (2013). Additionally, a time-space diagram for all buses operating on all routes has been developed, and bus speed profiles are obtained using the methodology proposed by Cortés et al. (2011). Using these processes, a detailed database of trips that contains a boarding stop, an alighting stop, a sequence of routes taken, travel time, transfer time and the waiting time at transfer points can be obtained. For buses, the information generated includes the detailed trajectory of each bus along its route, speed profiles and the load profiles per route.

These estimations have been validated by Munizaga et al. (2014) using exogenous data from measurements, surveys and personal interviews. Although the validation is positive, showing a correct estimation for over 80% of the cases, there are a significant number of cases in which the estimation is not correct. Therefore, for the estimation of level-of-service indicators, it is important to apply filters and corrections, which ensure that the real travel experience is well represented. It is particularly relevant to verify all of the extreme cases as well, separating a poor travel experience from a wrongly estimated trip. A detailed analysis of suspicious cases was performed by Núñez (2015).

Using the filtered information, the objective of developing global indicators to monitor system performance is investigated. From the information available, we calculate the travel time, number of stages per trip, travel speed and distance. These indicators can be computed at different levels of aggregation.

## 3.1 Global Indicators

The global system indicators calculated for data corresponding to a week in May 2014 are presented in Table 1. The table shows that the average values are quite reasonable for a city of 6.6 million people, showing an average travel time that is marginally above 30 minutes and 1.4 stages per trip. Considering peak-use hours only, the average travel times are marginally higher, particularly during the afternoon peak when the travel speed is lower. It must be noted that the travel time and the other indicators only consider the part of the trip that can be identified from the passive data available (i.e., they do not include the walking time both before the first stage at the origin and after the last stage at the destination). The waiting time at the origin is estimated from the observed headways of the bus line boarded or from the programmed headway if the first stage occurs in the metro.

**Table 1.** Global indicators in May 2014

| | Trips [per/day] | Travel Time [min] | Stages per Trip | Speed [km/h] | Distance [km] |
|---|---|---|---|---|---|
| Day | 4,002,525 | 31.3 | 1.4 | 14.7 | 7.7 |
| Morning Peak | 415,090 | 32.9 | 1.4 | 15.0 | 8.3 |
| Afternoon Peak | 401,941 | 34.5 | 1.4 | 14.1 | 8.1 |

To examine these values with more disaggregation in time, Figure 1 shows different indicators for 30-minute periods constructed based on the time of the first validation. The data shows that there is a significant peak in the travel time, number of stages and trip distance for trips initiating at approximately 5AM, which is significantly earlier than the morning peak hour in terms of the number of trips, which occurs near 7:30AM. This shows a concentration of trips with poor travel conditions in terms of the distance and the number of stages required to complete the trip in the early hours of the day; this finding is likely caused by those users that travel in these poor travel conditions being forced to depart earlier to arrive at their work or study locations on time. There is no such effect during the afternoon peak because the time concentration for the return trip depends on the location of the activities that the users are conducting and not on their residential locations. Those long trips likely occur in the afternoon as well but are not apparent in the average values presented at this level of aggregation.

## 3.2 Indicators at the Municipality Level

The Santiago metropolitan area contains 37 municipalities, each of which is well defined in terms of its area; several sources of information are available at this level. Therefore, municipalities are a natural level of disaggregation for analysis. We consider this level to be appropriate to calculate indicators related to the spatial structure of trips on the public

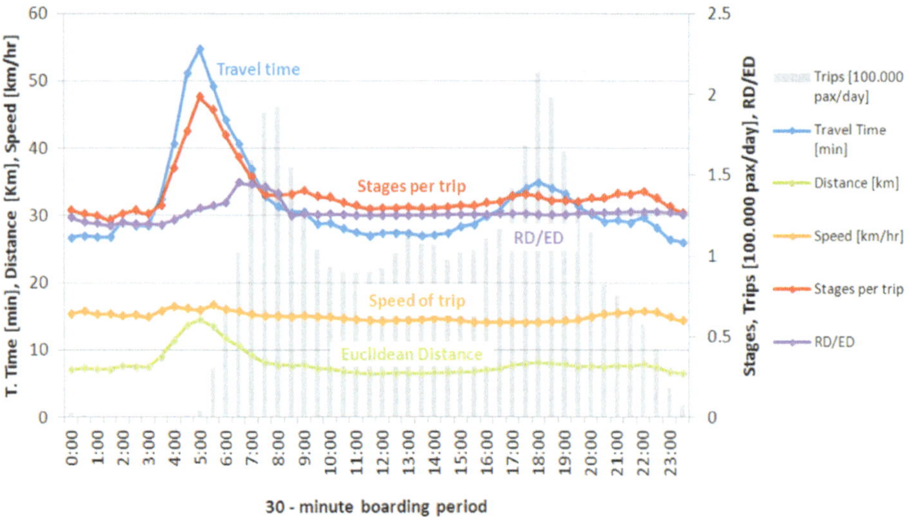


**Fig. 1.** Trips, stages, travel time, distance, speed and RD/ED

transportation system. In Figure 2, we present two variables (i.e., travel speed and route directness), where the route directness is calculated by the route distance divided by the Euclidean distance. In this graph, each circle represents a pair of municipalities, and the circle area is proportional to the number of trips observed in that OD pair. This particular graph was developed for the trips that arrive at their destinations between 8AM and 9AM. The data shows that the average travel speed varies significantly depending on the municipality where the trip originated, ranging from values below 10 km/h to values above 25 km/h on average. In addition,

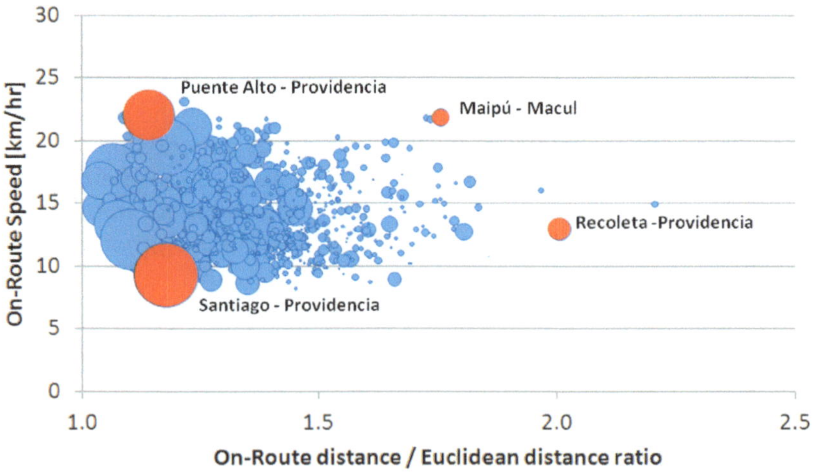


**Fig. 2.** On-route speed vs. route directness

large differences are observed in terms of route directness, ranging from values near one to values above two. The combination of both effects is shown in cases such as the route from Recoleta to Providencia, where users face a low speed of travel and indirect routes. On the other extreme, the route from Puente Alto to Providencia exhibits one of the highest average travel speeds and direct routes. Intermediate cases can have direct routes but low speed (e.g., Santiago-Providencia) or high speed but indirect routes (e.g., Maipú-Macul).

### 3.3 Indicators at Zone Level

To analyse the spatial effects of this system in more detail, a more disaggregated level must be used. For Santiago, we used the zoning that the Santiago transit authority (DTPM) uses for most of the analysis that they perform. The zoning, which is called "777", has approximately 800 zones and is compatible with the municipality zoning. At this level of aggregation, it is not possible to visualize the full OD structure; therefore, we focused on particular destinations to provide a few examples: the Santiago CBD (Santiago Centro); two zones with a mixture of activities that are residential areas and concentrated business and commercial activities (Providencia and Las Condes); and a high income residential zone (Lo Barnechea) that is located at the north eastern end of the city, which is a destination of many domestic work trips. Figure 3 shows the average travel time from each zone to Santiago Centro, Providencia, Las Condes and Lo Barnechea during the morning peak-use hours. Metro lines are outlined in red to illustrate the effect of certain important public transport infrastructure elements. The travel time intervals are shown by a colour code where blue is less than 30 minutes, and red is over 75 minutes. The values represented in that figure are the average values of the observed trips, which were calculated with at least five observations. A grey colour is used to identify cases in which information is not available (i.e., less than five observations).

The figures show that the city centre is not equally accessible from all possible origins. The presence of metro lines is associated with faster access to the city centre. It is also apparent that the relation between travel time and distance is not direct. Considering the other destination zones, Providencia and Las Condes can be reached in less than 30 minutes only from nearby zones and from zones associated with a metro line; additionally, few trips are shown with a destination in Lo Barnechea during the morning peak-use hours, and most require over 75 minutes. The trips with nearby origins can be made in less than one hour. This type of analysis can be performed for any location/period of interest.

A similar analysis can be performed with the number of trip stages required to reach any specific destination at any time period. Additionally, this information can be compared between different periods of time.

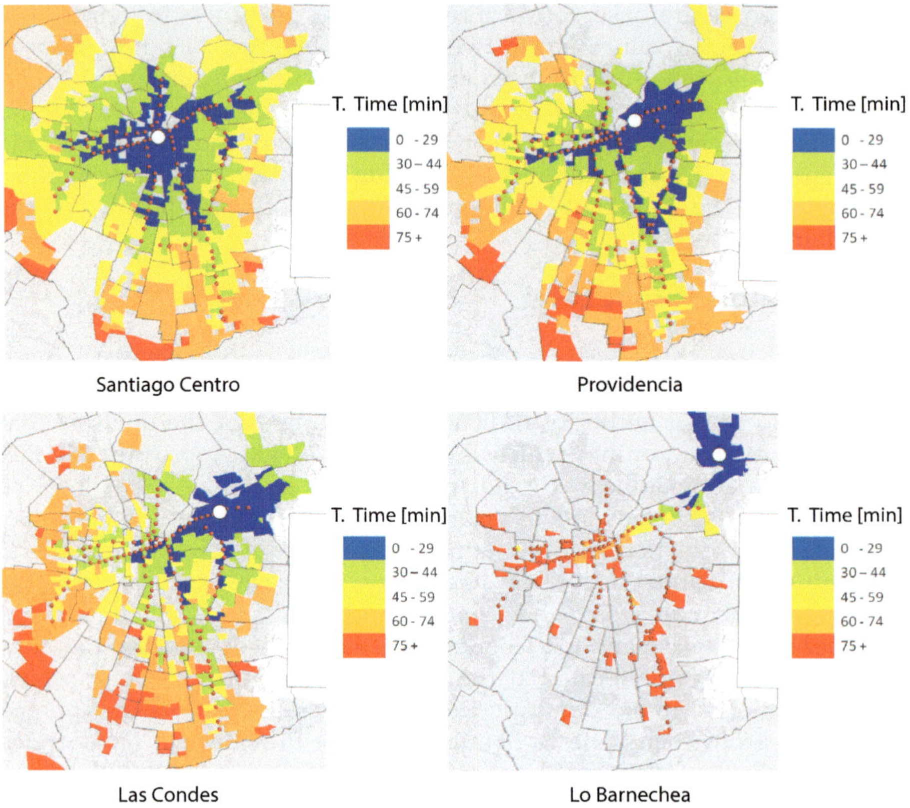


**Fig. 3.** Travel times to access Santiago Centro, Providencia, Las Condes and Lo Barnechea from any origin zone, May 2014, morning peak-use hours

Figure 4 shows the variation in the number of trip stages required to reach the city centre between 2013 and 2014. A negative difference represents a decrease in the number of trip stages, and a positive difference represents an increase. The colour code used ranges from blue for a large decrease to red for a large increase; yellow represents a mild change, either positive or negative, or none; white lines represent bus corridors with a segregated right of way; and black lines indicate the bus lines that move through the CBD. Figure 4 shows that there have been moderate or no changes in most of the zones (i.e., yellow dominates); however, there is a significant group of zones in the north western and western parts of the city where the number of trip stages required to reach the city centre has increased and, in some cases, significantly. This increase is explained by the closure of an important street in the city centre, which forced detours for all bus routes that used that street to travel outside of the CBD. Because all routes in the west and northwest connected the CBD using that street, passengers that could reach the central area without transfers in 2013 are now forced to make a transfer. This example shows that these indicators can be sensitive to changes in the system and ultimately to the behaviour of users.

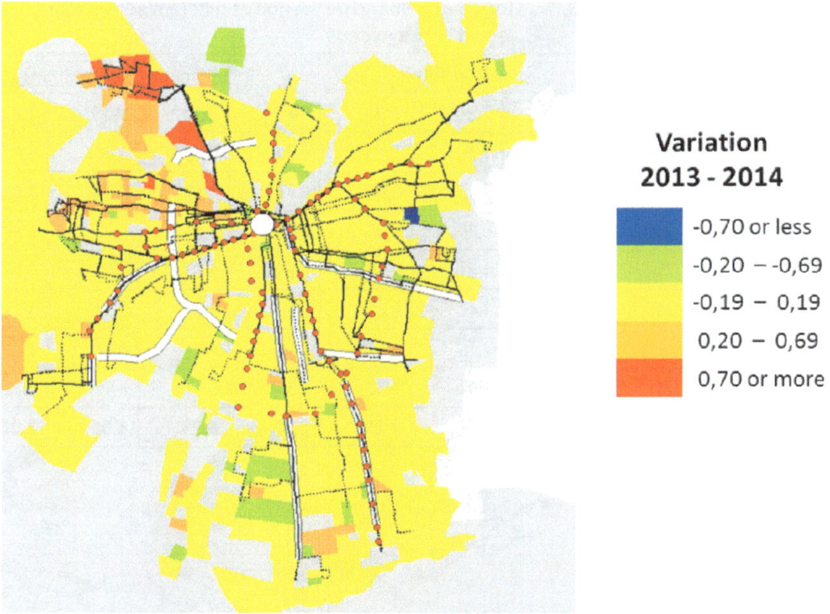


**Fig. 4.** Variation in average stages per trip to the CBD between April 2013 and May 2014

## 3.4 Indicators at the Avenue Level

At the avenue level, bus flows, passenger flows and the speeds of buses can be observed. To compute these variables, all bus routes that use a specific corridor can be identified, and the GPS pulses, which are projected onto the path, are used to develop a time-space grid representation. In this representation, space is included as a linear change along the road. Segments of different lengths are defined such that traffic and demand conditions are homogeneous (Gibson et al. 2015). This representation allows for the calculation of bus flows and travel times, which are then aggregated into the average commercial speed using the methodology proposed by Cortés et al. (2011). Additionally, incorporating smart card transactions and their alighting-bus-stop estimations, we are able to obtain the number of boarding and alighting passengers per stop and the number of times buses stop at bus stops. Table 2 and Figure 5 show these values for Santa Rosa Avenue, where segments 3 through 5 correspond to a bus corridor with a longitudinally, physically separated right-of-way, as defined by Vuchic (2007). In the remainder of the segments shown, buses share the road with private cars (i.e., mixed traffic). Table 2 shows bus flows by segment, stops per km travelled by a bus and the summation of boarding and alighting passengers (i.e., demand). Figure 5 shows the average commercial speed by road segment and time period. The difference between segregated bus way segments and mixed traffic segments is clearly shown.

**Table 2.** Average bus flows, stops and demand per period at different sections of Santa Rosa Avenue

| Section \ Period | Flow (bus/h) | | | Stop/bus-km | | | Demand (pass/bus-km) | | |
|---|---|---|---|---|---|---|---|---|---|
| | 7.30-9 | 9-19 | 19-21 | 7.30-9 | 9-19 | 19-21 | 7.30-9 | 9-19 | 19-21 |
| 1 | 122 | 85 | 100 | 1.36 | 1.05 | 0.79 | 4.71 | 2.66 | 2.07 |
| 2 | 121 | 85 | 101 | 1.39 | 0.97 | 0.73 | 4.51 | 2.11 | 1.69 |
| 3 | 126 | 90 | 107 | 0.89 | 0.63 | 0.47 | 2.40 | 1.20 | 0.88 |
| 4 | 111 | 78 | 94 | 1.53 | 1.09 | 0.85 | 6.10 | 2.68 | 2.23 |
| 5 | 110 | 78 | 94 | 1.47 | 0.85 | 0.59 | 3.93 | 1.57 | 1.13 |
| 6 | 108 | 78 | 93 | 2.16 | 1.57 | 0.90 | 8.65 | 4.19 | 2.19 |
| 7 | 101 | 74 | 88 | 1.91 | 1.51 | 1.44 | 13.4 | 6.99 | 4.12 |
| **Total** | **114** | **81** | **97** | **1.53** | **1.09** | **0.78** | **6.25** | **3.06** | **2.05** |

*Source*: Gibson et al., 2015.

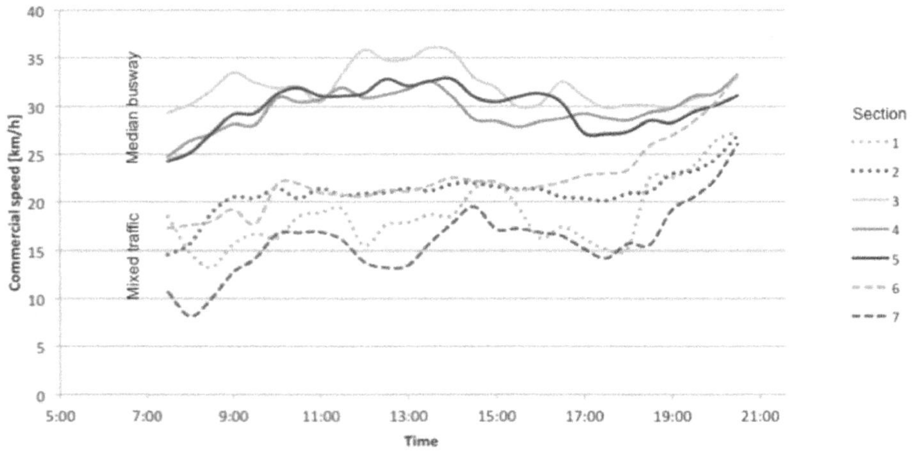


**Fig. 5.** Average commercial speed per section on Santa Rose Avenue

*Source*: Gibson et al., 2015.

## 3.5 Bus-stop-level Indicators

The relevant indicators at the bus-stop level are those related to bus flow and demand levels. Using the aforementioned time-space diagram, an interpolation procedure can be applied to estimate the instant when each bus passes through a bus stop, thus allowing the calculation of headways between buses and their variability. Additionally, for passengers boarding buses, a boarding bus stop is identified using the estimation of the instant when the bus was at the bus stop and of the validation time. The alighting stop or station is estimated using the method proposed by Munizaga and Palma (2012), and the locations where activities are conducted are identified using the activity detection method proposed by Devillaine et al. (2013). This allows the computation of boarding and alighting flows by bus stop, bus station and metro station. Figure 6 shows two screenshots of a self-made visualization tool that permits the selection of different types of trip per day, mode, number of stages, and time to describe the origin-destination structure. The selected trips are represented with their origin in the left side, and destination in the right side, using two identical maps. Yellow and purple circles are used to represent origin and destinations flows, using a proportional representation of volume.

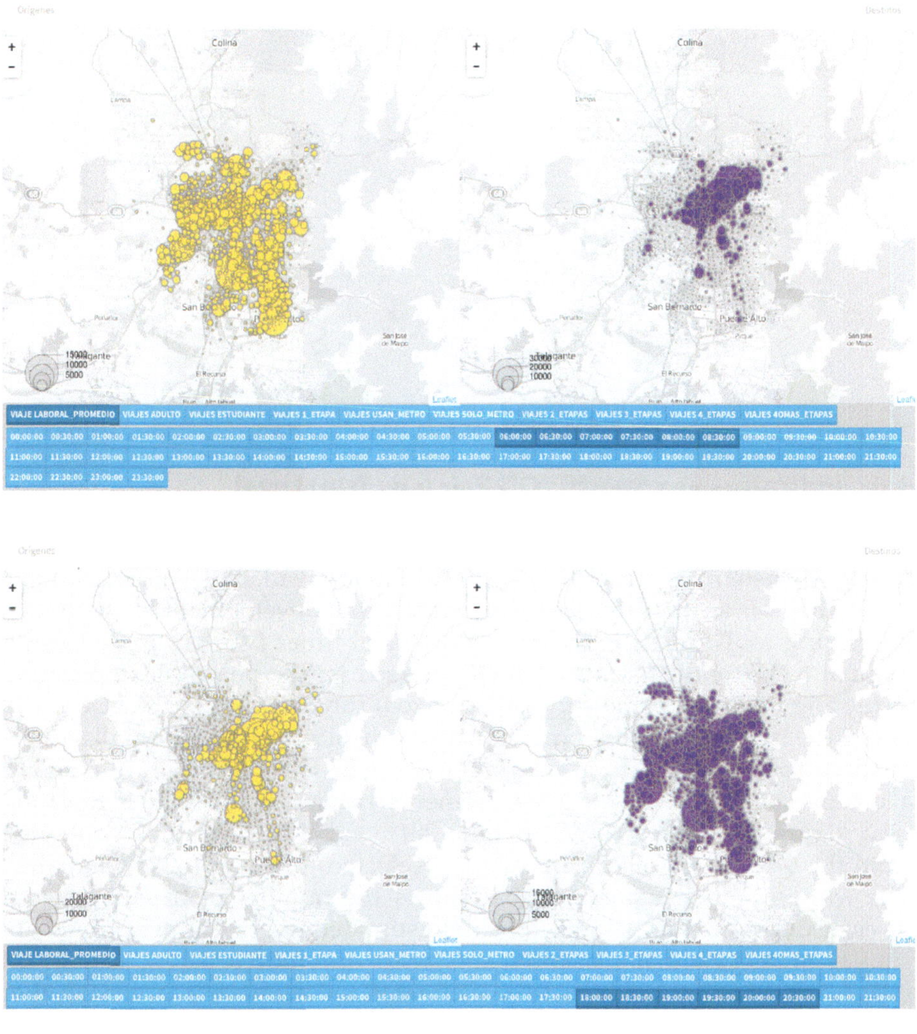


**Fig. 6.** Visual representation of the origin-destination flows

Figure 7 shows a histogram of the waiting times calculated at each of the more than 11,000 bus stops. The line shows the accumulated percentage (secondary scale) at each level of the histogram. In this figure, the waiting time is calculated from the headway observed (i.e., the time between the bus boarded by the user and the previous bus of the same line at that stop). Given that headway, a random value between zero and the headway is assumed to be the waiting time for this passenger.

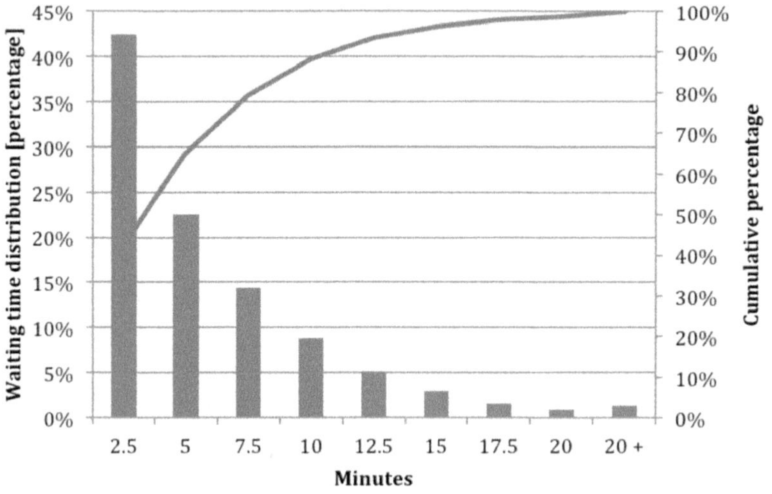


**Fig. 7.** Bus waiting times in the first trip stage

## 3.6 Indicators at a Specific OD Pair (i.e., Trip) Level

One important variable to measure the quality of service in public transport systems is travel time. From the data available, we can estimate travel time for any specific OD pair. Prior to the development of these data processing tools, certain monitoring of travel times was performed using manual measurements in certain OD pairs from the boarding stop/station to the alighting one using a given (i.e., fixed) travel strategy (DICTUC 2011). Given the high cost of this type of measurement procedure, a small sample of three trips for each time interval and for 27 OD pairs. Figure 8 shows a comparison between the measurements recorded during the morning peak hours (i.e., 7 to 10 AM) of June 2011 and the travel times obtained from the OD matrix of April 2012 for a selected group of those OD pairs[1]. The white number inside each bar shows the number of trips registered in each OD pair. It is shown that the estimated average travel times are similar to the manual measurements with differences below 10% in all cases. Additionally, given that the estimated values have a larger number

[1] The selected OD pairs for comparison are those in which the travel strategy has not been significantly affected by route changes, and a large number of observations is available.

of observations, their variance can also be calculated; this was not possible with the three manual measurements. The data shows that the variability is significantly different between these OD pairs. It is also worth recalling that a large variability in travel times negatively affects the perceived quality of service.

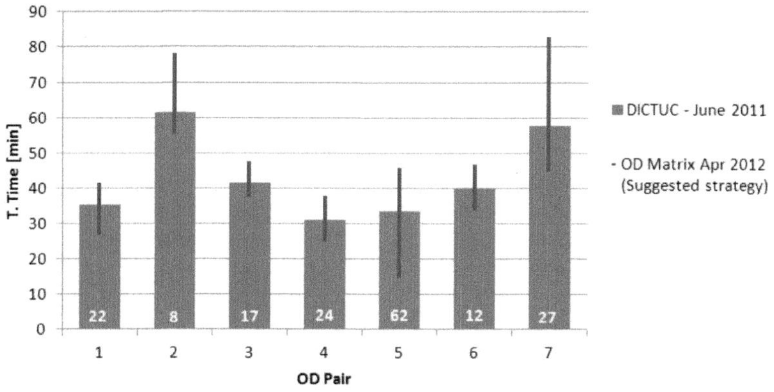


**Fig. 8.** Comparison of manual measurements and estimated travel times

One advantage of the passive data used in this study is that it provides a large sample size, allowing the production of average figures and an analysis of their variability. An example of this is shown in Figure 9, where the average travel time for all trips of a specific length is shown with the corresponding travel time for the 5th and 95th percentiles. These results show that there can be significant variability behind the averages. As expected, the variability increases with distance in this case.

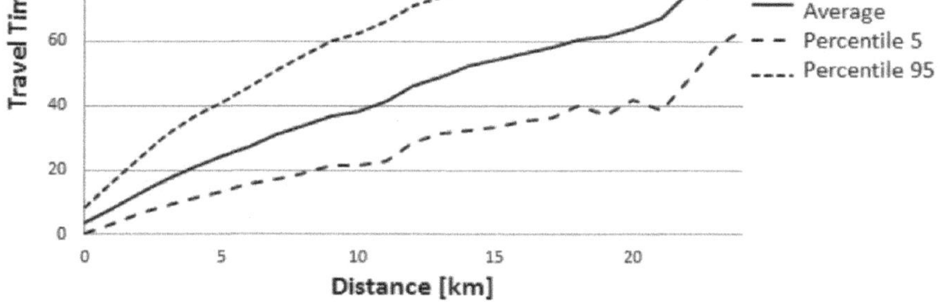


**Fig. 9.** Estimated travel time for different trip lengths and their variability

## 4. CONCLUDING REMARKS

This study presents tools that can be developed to monitor public transport systems, using information obtained from passive data. Detailed analyses can be performed using passive data and can contribute to the understanding of a system, its capacity and its performance at a low cost. Taking advantage of broad and precise temporal and spatial data, tools can be developed to monitor the system at different levels of aggregation to create and analyse average figures and the variability of phenomena of interest; thus, the proposed methods are critical to understand the quality of service of urban public transport systems.

At an aggregate level, global indicators provide a general idea of the system magnitude and its average behaviour as well as certain differences over time. At the strategic municipality level, the gross differences between zones can be observed; these can then be analysed in detail at the zone level. However, the difficulty of working with a larger number of zones must be addressed. Graphic analyses can be helpful for this purpose and the use of GIS-tools to show the geographical information over large number of zones has shown to ease the analyses. Moreover, animated views of the geographical data can enrich the understanding of the information. The challenge of creating innovative ways to present these data arises as a new issue by itself. If the infrastructure is of interest, the likely proper unit of analysis is the avenue, road or railway where buses or trains are operating. More detailed analyses can be performed at the bus stop or station level, where vehicles and passengers flow; time distributions can also be observed in high resolution. Finally, this type of data also allows the analysis of specific trips or OD pairs.

Using data from Transantiago, the public transport system of Santiago, Chile, this study has shown that it is possible to determine the level-of-service, mobility and accessibility indicators using passive data collected from automatic fare collection and vehicle location systems, complemented with data from geographic information systems. Significant effort is required in terms of the development of the methodologies, data management and the construction of the tools necessary to obtain the indicators from passive data; however, valuable results can be obtained from it and periodically replicated with new data at very low costs, if the methodologies and tools were already developed. A next step to improve the quality of this information is the inclusion of other passive data sources (e.g. smartphones) and some manual measurements (for those critical data that is impossible to obtain automatically) to complement the automatically collected data and enrich the information that can be obtained.

## ACKNOWLEDGEMENTS

This study was partially funded by Fondef (Grant D10I1002) and the Complex Engineering Systems Institute, Chile (Grants ICM P-05-004-F, CONICYT FBO16).

## REFERENCES

Amaya, M. and Munizaga, M.A. 2015. Smartcard data analysis 2.0: estimating the residence zone of frequent public transport users to analyse travel patterns. Working paper, Universidad de Chile.

Bagchi, M. and White, P.R. 2005 The potential of public transport smart card data. Transport Policy 12: 464-474.

Chakirov, A. and Erath, A. 2011a. Use of public transport smart card fare payment data for travel behaviour analysis in Singapore, paper presented at the 16th international conference of Hong Kong Society for Transportation Studies, Hong Kong, December.

Cortés, C., Gibson, J., Gschwender, A., Munizaga M. and Zúñiga, M. 2011. Comercial bus speed diagnosis based on gps-monitored data. Transportation Research Part C 19: 695-707.

Dell'Ollio, L., Ibeas, A. and Cecín, P. 2010. Modelling user's perception of bus transit quality. Transport Policy 17: 388-397.

Dell'Ollio, L., Ibeas, A., Cecín, P. and dell'Ollio, F. 2011. Willingness to pay for improving service quality in a multimodal area. Transportation Research Part C 19: 1060-1070.

Devillaine, F., Munizaga, M.A. and Trepanier, M. 2012. Detection of activities of public transport users by analysing smart card data. Transportation Research Record 2276: 48-55.

DICTUC. 2011. Elaboración de Indicadores de Desempeño del Sistema de Transporte Público, Etapa II, Orden de Trabajo Nº8. Coordinación Transantiago.

Donoso, P., Munizaga, M.A. and Rivera, J. 2013. Measuring user satisfaction in transport services: methodology and application. pp. 603-623 *In*: Zmud, J., M. Lee-Gosselin, M.A. Munizaga, and J.A. Carrasco [eds.] Transport Survey Methods: Best Practice for Decision Making. Emerald.

Eboli, L. and Mazzulla, G. 2012. Performance indicators for an objective measure of public transporte service quality. European Transport 51, Paper n°3.

Fielding, G.J., Babisky, T.T. and Brenner, M.E. 1985. Performance evaluation for bus transit. Transportation Research Part A 19: 73-82.

Gibson, J., Munizaga, M.A., Schneider, C. and Tirachini, A. 2015. Median busways versus mixed-traffic: estimation of bus travel time under different priority conditions with explicit modelling of delay at traffic signals. Transportation Research Board 94th Annual Meeting, Washington DC. Accesible at http://docs.trb.org/prp/15-1260.pdf

Hensher, D., Stopher, P. and Bullock, P. 2003. Service quality – developing a service quality index in the provision of commercial bus contracts. Transportation Research Part A 37: 499-517.

Jang, W. 2010. Travel time and transfer analysis using transit Smart card data. Transportation Research Record 2144: 142-149.

Lee, S.G., and Hickman, M. 2011. Travel pattern analysis using smart card data of regular users. Preprint CD-ROM for the 90th Annual Meeting of Transportation Research Board of the National Academies, Washington, D.C., 2011.

Ma, X., Wu, Y., Wang, Y., Chen, F. and Liu, J. 2013. Mining smart card data for transit riders' travel patterns. Transportation Research Part C 36: 1-12.

Morency, C., Trépanier, M. and Agard, B. 2007 Measuring transit use variability with smart-card data. Transport Policy 14: 193-203.

Munizaga, M.A. and Palma, C. 2012. Estimation of a disaggregate multimodal public transport origin-destination matrix from passive Smart card data from Santiago, Chile. Transportation Research Part C 24: 9-18.

Munizaga, M.A., Devillaine, F., Navarrete, C. and Silva, D. 2014. Validating travel behavior estimated from smartcard data. Transportation Research Part C 44: 70-79.

Núñez, C. 2005. Cálculo de indicadores de calidad de servicio para el sistema de transporte público de Santiago a partir de datos pasivos. MSc thesis, Universidad de Chile.

Ortúzar, J. de D., A.M. Ivelic, A. and Candia, A. User perception of public transport level of service. pp. 123-142 *In*: Stopher, P.R. and M. Lee–Gosselin. [eds.] 1997 Understanding Travel Behavior in an Era of Change. Elsevier.

Park, J.Y., Kim, D.-J. and Y. Lim. 2008. Use of smart card data to define public transit use in Seoul, Korea. Transportation Research Record 2063: 3-9.

Raveau, S., Guo, Z., Muñoz, J.C. and Wilson, N.H.M. 2014. A behavioral comparison of route choice on metro networks: Time, transfers, crowding, topology and socio – demographics. Transportation Research Part A 66: 185-195.

Seaborn, C., Attanucci, J. and Wilson, N.H.M. 2009. Analyzing multimodal public transport journeys in London with smart card fare payment data. Transportation Research Record 2121: 55-62.

TRB. 2013. Transit Capacity and and Quality of Service Manual, Third Edition. TCRP Report 165, Transportation Research Board, Washington DC.

Trépanier, M., Tranchant, N. and Chapleau, R. 2007. Individual trip destination estimation in a transit smart card automated fare collection system. Journal of Intelligent Transportation Systems 11: 1-14.

Tyrinopoulos, Y. and Antoniou, C. 2008. Public transport user satisfaction: variability and policy implications. Transport Policy 15: 260-272.

Vuchic, V.R. 2007. Urban Transit: Systems and Technology. John Wiley & Sons, New Jersey, USA.

Yáñez, M.F., Raveau, S. and de D. Ortúzar, J. 2010. Inclusion of latent variables in Mixed Logit models: Modelling and forecasting. Transportation Research Part A 44: 744-753.

Zhao, J., Rahbee, A. and Wilson, N. 2007. Estimating a rail passenger trip origin-destination matrix using automatic data collection systems. Computer-Aided Civil and Infrastructure Engineering 22: 376-387.

Utsunomiya, M., Attanucci, J. and Wilson, N. 2006. Potential uses of transit smart card registration and transaction data to improve transit planning. Transportation Research Record 1971: 119-126.

## AUTHORS BIOGRAPHY

**Marcela A. Munizaga** is Associate Professor at Universidad de Chile. Specialist on transport demand modeling, survey methods, and data processing methods. In the last few years has leaded research on smartcard data and developed applications to obtain valuable information from automatically generated databases, which have been transferred to practice and used for planning purposes. Currently leads a research group in Smartcities, is Co-Chair of the International Steering Committee for Transport Survey Conferences ISCTSC and Co-Editor of the Spanish language Journal Revista Ingeniería de Transporte.

**César Núñez** (born in Santiago, Chile, 1987) is a Civil Engineer and MSc in Transport Science at Universidad de Chile, where he earned his degree with honors. Since 2013, he works as a research and development engineer at the public transport authority in Santiago, Chile, where one of his main tasks is the analysis of smartcard and GPS databases. In addition, he collaborates as a teaching assistant in the Public Transport Planning course at Universidad de Chile.

**Antonio Gschwender**, born in 1972, is a Civil Engineer, Magister in Transport Science (Universidad de Chile, 2000) and PhD in Engineering (Bergische Universität Wuppertal, Germany, 2007). He presently assigns his time to professional work, academic work, and being a husband and a father. He works at the public transport authority in Santiago, Chile, and teaches Public Transport Planning at Universidad de Chile, where he collaborates in research as well.

Chapter 10

# Evaluation of Bus Service Key Performance Indicators using Smart Card Data

**M. Trépanier* and C. Morency**[1]

**ABSTRACT**

Whilst smart card systems are becoming the standards among transit authorities, there is a growing need to valorize their collected data. Smart card data provides a continuous stream of passenger transactions that could be used to derive key performance indicators (KPI). This chapter presents methods used to calculate KPIs such as commercial speed, vehicle-kilometres, passenger-kilometres, schedule adherence and fare evasion for Canadian public transit networks. It shows the advantages of using smart card over traditional automated vehicle location systems (AVL) or automated passenger counting systems (APC), because with smart card KPI could be broken down by fare type or any other specific attributes linked to transactions. However, there are limitations to this approach because not all passengers use a smart card. Hence, fusion of data is somehow needed to complete the KPI portrait.

## 1. INTRODUCTION

The primary role of smart card fare collection systems (or AFC, automated fare collection systems) in public transport agencies is to facilitate revenue management. The continuous collection of transaction data can also be seen as an indirect sight of performance of the public transport systems, for both supply and demand. As a matter of fact, the smart card system data reveals spatio-temporal information on users, vehicles and routes. These spatial-temporal observations could be used to derive, calculate and tabulate key performance indicators (KPI) of the public transport network such as commercial speed, vehicle-kilometres, passenger-kilometres, schedule adherence, etc.

[1] Polytechnique Montreal, 2500 chemin de Polytechnique, station Centre-Ville, Montreal, Quebec, Canada. Email: mtrepanier@polymtl.ca and Email: cmorency@polymtl.ca

* Corresponding author

This chapter presents KPI that could be calculated from smart card data with the help of innovative processing methods. It will also emphasize the advantages of using smart card over traditional automated vehicle location systems (AVL) and automated passenger counting systems (APC). The chapter begins with some key background elements on this topic. Then, presented a series of indicators; descried the estimation method as well as examples. The chapter concludes with a discussion on the challenges related to the continuous estimation of the proposed KPI, current limitations and research perspectives.

## 2. BACKGROUND

Since the pioneer work of Bagchi and White (2005), a lot of research has been done to value the use of smart card data in public transportation. In this review, it will focus on performance indicators and destination estimation, two elements that are essential to the work presented in this chapter. For a broader view of the potential of smart card data, please refer to Pelletier et al. 2011.

### 2.1 Performance Indicators

Through the years, many performance indicators became the standards across the public transit industry. The Transit Capacity and Quality of Service Manual (Kittelson and associates, 1999) uses six performance measures to evaluate the service: service frequency, hours of service, service coverage, passenger loading, reliability and transit vs. automobile travel time. There are many examples of transit performance evaluation using the TCQSM (among others, Perk and Foreman 2003; Caulfield and O'Mahony 2004). In many cases, the KPI use surveys or on-board counts that provide static figures instead of continuous measurements.

The technology has evolved with the advent of Automated Vehicle Location systems (AVL). In these systems, vehicles are equipped with on-board GPS that record location as well as bidirectional telecommunications with a central server. This way, AVL can give a continuous evaluation of vehicle-specific indicators like commercial speed and schedule reliability. If the system is coupled to an automated passenger counting system (APC), more could be done to evaluate load profile and ultimately better manage the bus fleet and the service provision (Gillen et al. 2001; Nurul Hassan et al. 2013). Nowadays, real-time AVL data can be used to provide adjusted schedules to the passengers through Passenger Information Systems (PIS), even for smaller transit authorities (Cachulo et al. 2012).

As demonstrated by Trépanier et al. (2009), KPI can also be obtained through the analysis of smart card data. Because smart cards could be used to follow the behaviour of passengers and associated to specific fare categories, KPI could be ventilated through passenger and fare types, or made specific to given segments (i.e., passengers that boarded on that part of the city, at this time, etc.). However, as in most smart card system, there is no personal information on travellers, although data fusion approaches could provide more insights on the personal attributes related to smart card transactions (Kusakabe and Asakura (2014) and Chapter 5 in this book).

## 2.2 Destination Estimation Algorithm

Many smart card automated fare collection systems collect the fare at vehicle entrance and stations ("tap-in") and not at the exit ("tap-out"). Because transactions are only recorded the entrance location, new methods had to develop to estimate the alighting (exiting) point of the user making his journey on the public transport network. Trépanier et al. (2007) proposed a method based on sequence of trips made by a single card during a day. Examine each card's transactions separately. For each boarding, retain the sequence of stops that follows as potential candidate for alighting (see Figure 1). Retain the stop that is the nearest to the next boarding made during the day as the estimated stop. In the case of the last alighting of the day, deduct the location looking at the first boarding of the next day.

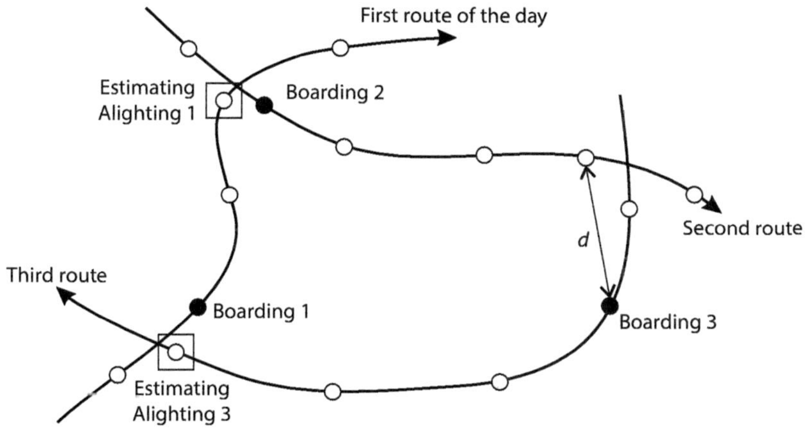


**Fig. 1.** Basic algorithm for destination estimation (From: Li and Trépanier 2015)

The method has been gradually improved through the years. Munizaga and Palma (2012) showed that these results can lead to production of very detailed origin-destination matrices. Recently, Li and Trépanier (2015) proposed an improved method for unlinked trips whose alighting cannot be estimated by the sequence-based method. Use a kernel density probabilistic method to find the alighting stop, looking at the historical data of the smart card.

## 3. INFORMATION SYSTEM

Assessment of public transport KPIs such as those presented thereafter necessitate the use of raw data collected in the smart card system. Data typically include:

- the identification and the date and timestamp of the transaction;
- the card number and, when available, the user number;
- the route and direction (if not subway) taken in the trip;
- the operational data like the bus number, vehicle type, assignment number, etc.;
- the location of the origin stop ("tap-in") and the location of the destination stop ("tap-out") or the estimated stop if not available;
- the fare type and any socio-demographic that would be available on users.

The examples given in this chapter were mainly calculated from three sources:

1. Data from the smart card automated fare collection system of the Société de transport de l'Outaouais (STO), a mid-size public transport authority in Gatineau, Québec, Canada (220,000 inhabitants, 300 buses), for a total of 65 million transactions over a 9-year period (from 2001 to 2010).
2. Data from the OPUS smart card system of the Commission interrégionale de transport des Laurentides (CITL), a suburb operator located to the northwestern part of Montréal, Québec, Canada (372,000 inhabitants).
3. Data from the OPUS smart card system of the Société de transport de Montréal, a large public transport agency operating a 68 km subway network and more than 1,500 buses (2 million inhabitants).

## 4. KPI ASSESSMENT

### 4.1 Error Detection

Error detection is the first task done with smart card data, before attempting to calculate any indicator based on transactional data. As for all enterprise information systems, smart card systems will contain a certain amount of systematic and random errors. Systematic errors could be caused by equipment malfunction (on-board device, card reader, wrong time clock, etc.) or by having the wrong vehicle assignment status (programmed vehicle go on a route but will service another), last minute changes, etc. Random errors are rarer but can result from a bad data manipulation, or wrong database command.

Errors could be detected through indicators based on logic according to the normal functioning of the public transit network. Figure 2 presents a space-time diagram that compares the planned service to the transactions log in the case of a single vehicle. The figure shows that at three places, the transactions are erroneous because the vehicle is not moving for long periods of time, but passengers are still boarding. In this case it is caused by a wrong vehicle assignment. The on-board smart card reader is then unable to find the correct stop location so the transactions are all assigned to the same stop.

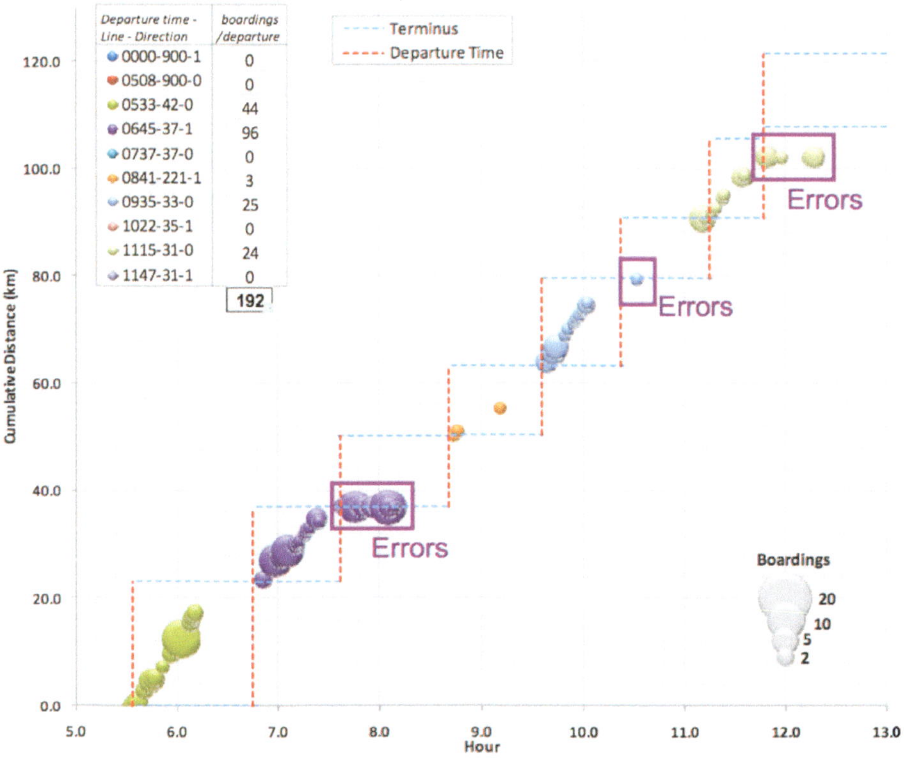


**Fig. 2.** Comparison between the attributed run of transactions and the planned service in a vehicle block with a time-space bubble diagram (From: Chu et al. 2009)

## 4.2 KPI Calculation Framework

The KPI calculation is based on the thorough examination of the smart card transactions made on individual vehicle runs. Figure 3 illustrates the conceptual framework used for the examples shown in the next section. The figure shows a space-time diagram of the bus passages on a single route. Plot the curves using sequence of smart card transactions. Represent each bus stop (where transactions occur) by half circles. The left half circle shows the number of boarding transactions, while the right one shows the number of alighting (exit) transactions. Calculate logically, the on-board load at a given stop by summing the boarding transactions and subtracting the alighting transactions before that stop. Please keep this figure in mind for the explanations in the next section.

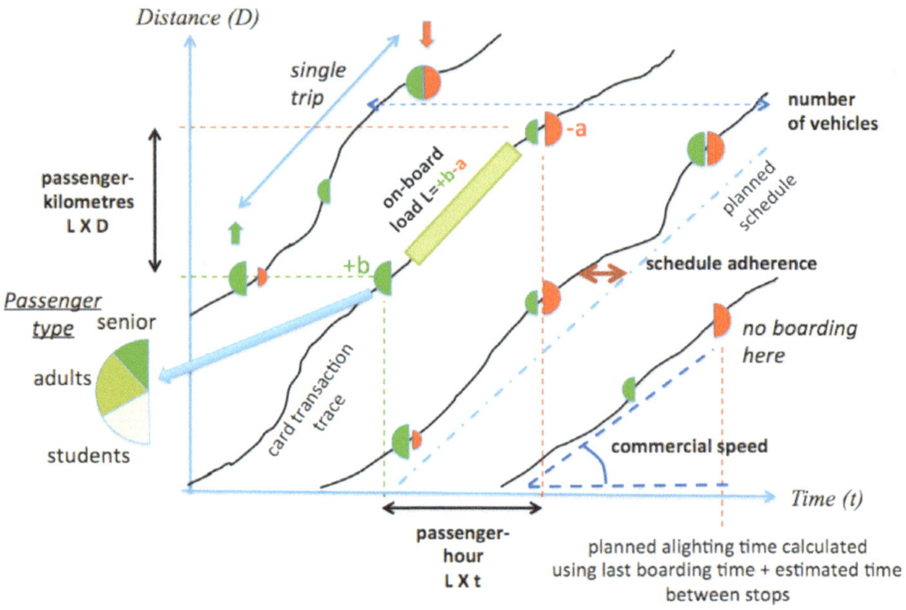


**Fig. 3.** Conceptual framework for KPI calculation

## 5. SOME EXAMPLES

This section presents some examples of KPI calculated from smart card data. The method refers to Figure 3.

### 5.1 Commercial Speed and Average Trip Distance and Duration

The commercial speed is the slope of the curve made by sequence of transactions along a route. The speed can be calculated using numerous smart card transaction timestamps available at different locations. The average trip distance and duration can also be calculated because each card could be followed from its boarding to its alighting. In "tap-in only" systems, the alighting time could be derived using the time of boarding located at the alighting. If there is no boarding taking place at this location, it uses the last known boarding time plus the estimated amount of time between the last known boarding location and the current alighting location. Table 1 presents some results from the Gatineau smart card system. The average speed varies on weekdays and, as expected, is higher on weekends. The average trip distance correlates this, which is longer on weekends for about the same trip duration.

**Table 1.** Key facts regarding transit supply during the four weeks of November 2006 by day of travel (From: Trépanier et al. 2009)

| Weekday | Average Speed (km/h) | Average Trip Distance (km) | Average Trip Duration (min.) |
|---|---|---|---|
| Sunday | 17.4 | 11.1 | 38.3 |
| Monday | 14.8 | 9.1 | 37.1 |
| Tuesday | 14.5 | 8.8 | 36.5 |
| Wednesday | 14.1 | 9.1 | 38.6 |
| Thursday | 13.9 | 9.0 | 38.9 |
| Friday | 13.7 | 9.0 | 39.6 |
| Saturday | 15.9 | 10.3 | 38.8 |

One may say that these indicators could be obtained from classical automated vehicle location systems (AVL). However, this is not the case. Calculate the average trip distances and durations looking at the individual behaviour is something that an AVL cannot do. The commercial speeds are also related to individual behaviour (when there is people aboard vehicles), which may slightly differ from commercial speed calculated from AVL.

## 5.2 Passenger-kilometres, Passenger-hours

The difference between KPI from smart card and KPI from AVL are better assessed when looking at passenger-kilometres and passenger-hours, common indicators used to measure the demand on a public transit network. Smart card data can be used to ventilate passengers by fare type, helping to measure the use of the network by types of users. Table 2 shows some KPI per card type for the STO network. It shows that express and interzone fares imply faster commercial speeds and longer trip lengths and durations. Smart card data allows the operator to know the amount of service consumed by users of different fares. In multi-operator schemes, this could help to share the revenues among authorities

**Table 2.** Key facts on demand by card type, November 2006 (From: Trépanier et al. 2009)

| Card Type | #Boarding | Pass-km | Pass-hr | Average Speed (km/h) | Average Length (km) | Average Duration (min.) |
|---|---|---|---|---|---|---|
| Adult-Regular | 46.2% | 38.9% | 42.3% | 17.5 | 7.0 | 24.1 |
| Adult-Express | 15.1% | 21.8% | 20.4% | 20.4 | 12.0 | 35.3 |
| Adult-Interzone | 3.0% | 11.5% | 8.4% | 26.2 | 31.8 | 72.9 |
| Student | 26.3% | 21.0% | 21.7% | 18.4 | 6.7 | 21.7 |
| Senior | 3.5% | 2.1% | 2.3% | 17.8 | 5.1 | 17.2 |
| Other | 6.0% | 4.7% | 5.0% | 17.9 | 6.6 | 22.0 |
| Total | 100% | 100% | 100% | 19.0 | 8.3 | 26.3 |

These figures can be examined longitudinally over time, as shown at Figure 4. Looking at average speed, the operator may know performance of the network or detect possible perturbations such as weather events. However, the use of KPI obtained by smart card data could be risky if the smart card is not used by a large part of the ridership, or if there are large usage discrepancies between routes, parts of the network or time periods. Sometimes, the smart card automated fare collection systems also collect cash payment and tickets. This is the case for the Montreal OPUS system, where it is possible to gather data from both transactions from smart cards and tickets/cash payments on board.

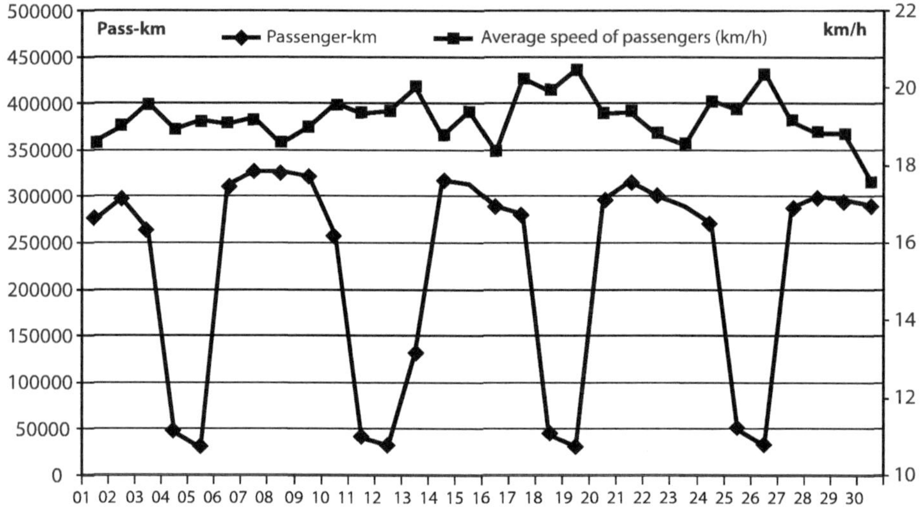


**Fig. 4.** Key figures for demand, per day, November 2006

## 5.3 Load Profile

Referring to Figure 3, the load aboard vehicles can be calculated by looking at the successive boarding and alighting of passengers. Because each transaction can be retrieved by fare type and other attributes, the load profile can be ventilated accordingly, supposing that the smart card is widely used by passengers. Figure 5 presents an example of load profile displayed interactively in the case of the STO network. The figure shows the stop-by-stop profile for a single bus run of a given day. With smart card systems, this profile can be generated for every single bus run, or can be aggregated or averaged with history, as needed by the operator. In this example, it can also query every stop to know boarding distribution and alighting passengers by fare type. There is also the possibility to filter the profile according to fare type. Following this logic, origin-destination matrices can be calculated for each route, for subsets of routes, or for the entire network.

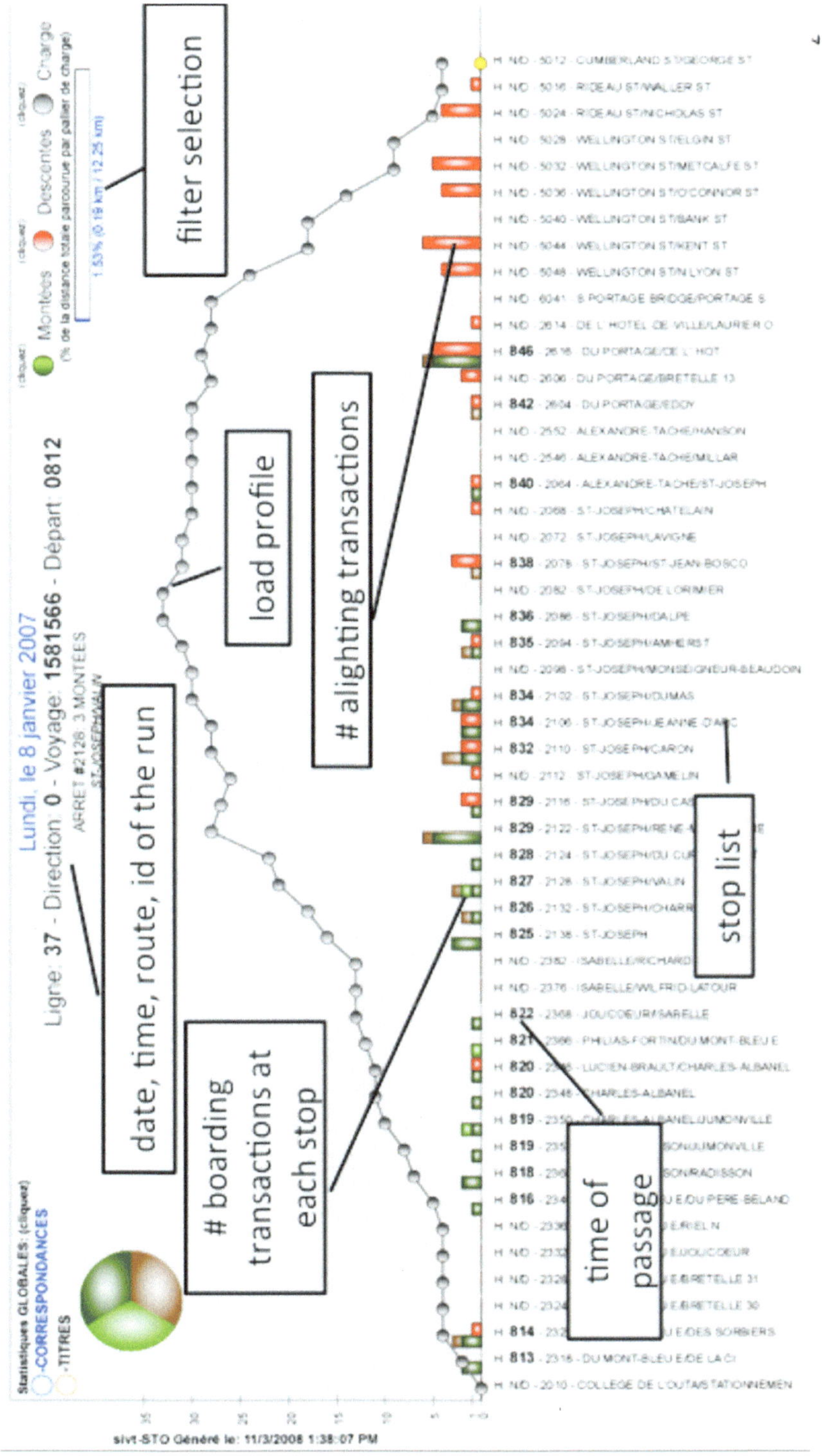


**Fig. 5.** An example of load profile obtained from smart card data (From: Trépanier and Vassivière 2008)

Load profiles are trickier to calculate for subway networks, because 1) transactions do not show the direction, or the vehicle taken at the boarding location and 2) passengers may have many path choices to reach their destination (alighting transaction). Si et al. (2014) proposed a destination-estimation method based on travel time, travel distance and the number of transfers for the Beijing subway network. They categorize users based on their occupation, salary and purpose of travel and their destination-estimation accuracy is above 85%. Sun and Schonfeld (2014) tried to find the exact time of boarding and alighting by looking at three factors: 1) the time between the transaction and boarding, 2) missing vehicle possibility and 3) the schedule for the subway and the theoretical transfer time at stations. The subway path choice question has also been thoroughly examined by Raveau et al. (2011) (also see Chapter 4 of this book).

## 5.4 Service Variability

As for AVL systems, smart card transactions show the service variability that can occur due to the traffic congestion or the passenger load aboard. Figure 6 shows the space-time diagram of a single route of the STO network during one day of January 2007. It shows the higher density of service at peak hours, but also bus bunching occurring towards the end of the route. The on-board load is also displayed on the chart.

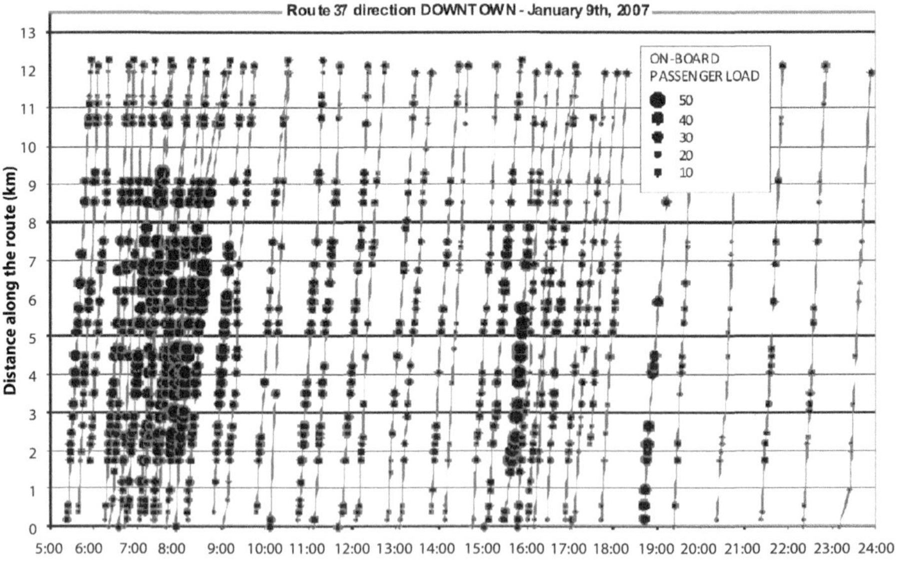


**Fig. 6.** Space-time diagram for route 37, direction DOWNTOWN, on January 9th 2007 (From: Trépanier and Vassivière 2008)

## 5.5 Service Fit

The load profile and the passenger-kilometres indicators can be joined to illustrate the quality of the service fit. Figure 7 presents the daily statistics for a single route of the STO network during the month of January 2007. The figure shows large variations in the number of passenger-kilometres throughout days, especially from a Monday to another, showing the gradual reprisal of the service after the Christmas and New Year Day Holidays. These variations might be put in perspectives and compared to the schedules that are usually stable during weekdays.

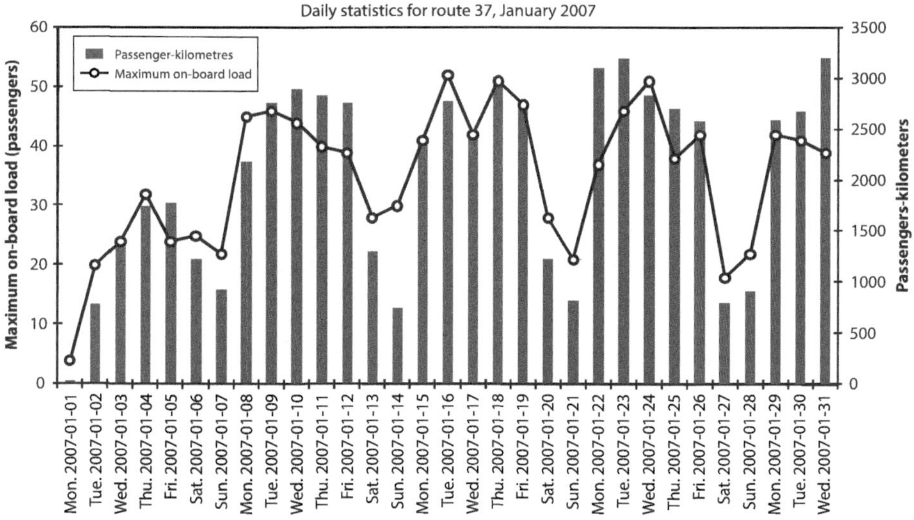


**Fig. 7.** Daily statistics for route 37, January 2007 (From: Trépanier et al. 2009)

Smart card data can be used to indirectly calculate the number of buses in service during the day, because the bus identification number is typically available in transaction data. This allows comparing the ridership to the number of vehicles as in Figure 8, helping to adjust the shoulders of the peak hours.

## 5.6 Schedule Adherence

Another interesting indicator that can be derived from smart card transaction data is schedule adherence. The respect of the schedule can be examined thoroughly, stop by stop, by comparing the transaction timestamps to the scheduled time of vehicle passages at stops. Schedule adherence can also be calculated without any prior knowledge of the schedule, or the network geometry. Figure 9 shows distribution of the differences between smart card timestamps and the "deducted" schedule of a single route of the STO network. In this case, the schedule is defined by

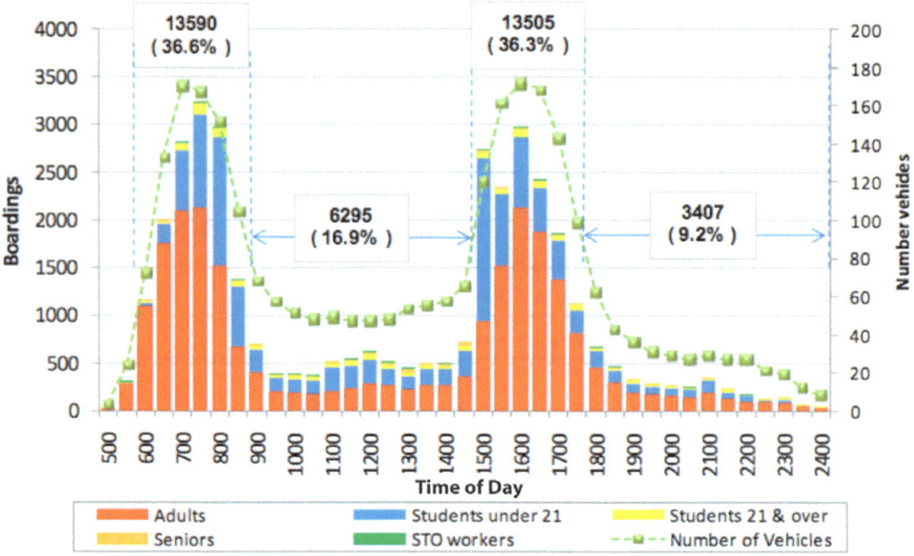


**Fig. 8.** Boarding by fare type and the estimated number of vehicles in service on a weekday (From: Chu et al. 2009)

the average of transaction timestamps by bus run and bus stop. The figure shows that the majority of timestamps are 0 to 2 minutes after the schedule, which is acceptable for a transit network, while passage ahead of the schedule or too late, might be avoided. Like other KPI, this type of analysis could be done for different fare types, customer groups, or other attributes available.

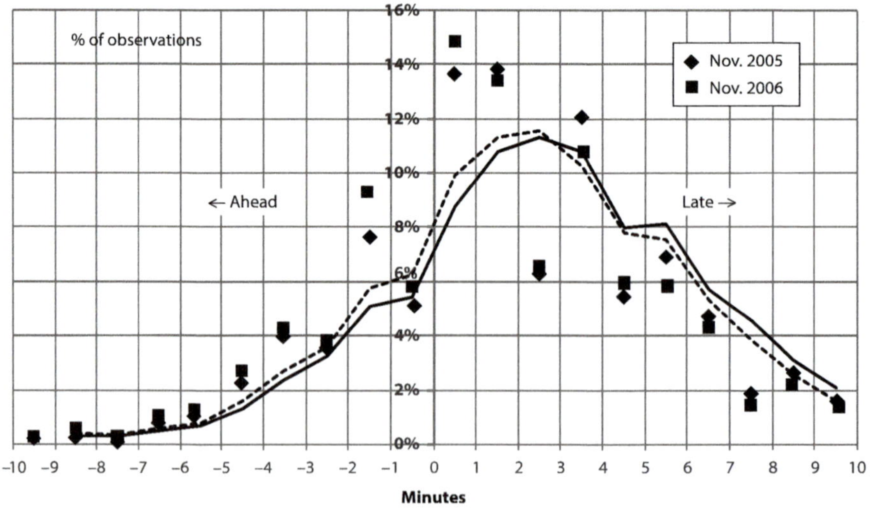


**Fig. 9.** Schedule adherence on route 37 based on November 2005 and November 2006 observations (From: Trépanier et al. 2009)

## 5.7 Fare Evasion

Pourmonet et al. (2015) proposed a fare evasion detection method for the public transport network of the Société de transport de Montréal. In Montréal, because the OPUS smart card system collects all transactions (cards, tickets, cash), it is possible to compare the "universe" of boarding transactions to the passenger counting done by the automated passenger counting system (APC), given that 1) there might be counting errors in both systems and 2) discrepancies between the systems might not be related to fare evasion, but could be a good indicator of it. The data processing method calculates a ratio between the smart card (SC) system count and the APC count. Figure 10 presents this SC-to-APC ratio for the hours of the day in April and October 2014 data (more than 10 million smart card transactions for both months). When the ratio is above 1, it means that the number of passengers from the smart card system is higher than the estimate taken using APC data (not likely to happen or might be due to the margin of error of APC. The figure shows that the ratio decreases during the day, showing a possible increase in fare evasion from morning to evening. However, the October figures seem a little better during midday. Of course, this indicator could be analysed for each route or stop and could enhance fare inspection by showing the locations and time where the phenomenon is more frequent.

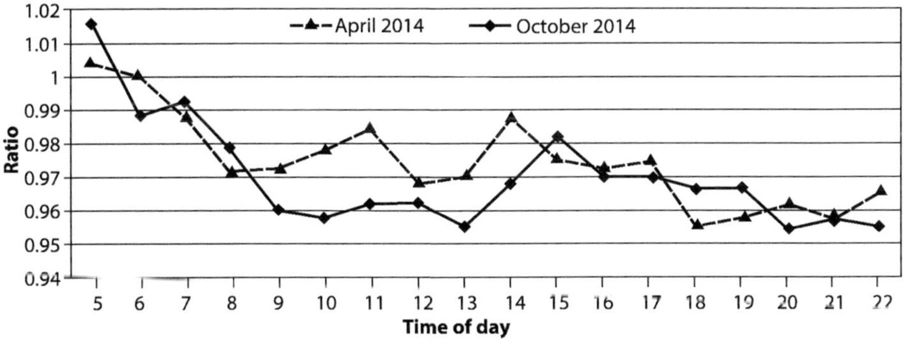


**Figu. 10.** APC-to-SC ratio for hours of the day, STM data (From: Pourmonet et al. 2015)

## 6. CONCLUSION

This chapter has presented a series of key performance indicators of public transport networks that could be calculated from smart card data: load profile, commercial speed, passenger-kilometre and passenger-hour, schedule adherence, fare evasion, service fit, etc. As shown, smart card data could be used to calculate the classical KPIs obtained from AVL and APC systems. But smart cards will also provide attributes like fare type, bringing new dimensions to the analysis of these KPIs.

## 6.1 Limitations and Challenges

Two limitations arise from these studies. First, all passengers in public transit networks do not use smart card. This may affect the representativeness of the KPIs calculated; still some indicators might be completed with APC data, if available. The second limitation is calculation related of these KPIs on a continuous basis. The information system of the transit authorities might be sufficiently mature to support the undergoing tasks of smart card data processing for errors, destination estimation and KPI calculation. The display, the processing and interpretation of this continuous KPI availability need enhancement and integrated into the public transit authority processes.

## 6.2 Perspectives

In the near future, the smart card integration, real-time APC, network geometry and schedule data will permit to better model passenger behaviour and the impacts of service changes on ridership. This "micro elasticity" of passenger behaviour will need a detailed examination of individual habits, plus a thorough analysis of the service quality at specific time and location, all put into context in the integrated urban mobility system. It will then be possible to predict the impacts of changes or perturbations over networks, helping to adjust the service to customer needs.

## ACKNOWLEDGEMENTS

The authors wish to acknowledge the supporters of these studies, which are the *Société de Transport de l'Outaouais*, the *Réseau de transport de Longueuil*, the *Société de transport de Montréal*, the *Agence métropolitaine de Montréal*, the Natural Science and Engineering Research Council of Canada (project RDCPJ 446107-12) and Thalès Research and Technologies.

## REFERENCES

Bagchi, M. and White, P.R. 2005. The potential of public transport smart card data. *Transport Policy*, 12 (5), September 2005, pp. 464-474.

Bertini, R. and El-Geneidy, A. 2003. Generating Transit Performance Measures with Archived Data. *Transportation Research Record*, 1841, pp. 109-119.

Bonneau, W. and eds. 2002. The role of smart cards in mass transit systems. *Card Technology Today*, June 2002, p. 10.

Cachulo, L., Rabadão, C., Fernandes, T., Perdigoto, F. and Faria, S. 2012. Real-Time Information System for Small and Medium Bus Operators. *Procedia Technology*, Volume 5, pp. 455-461.

Caulfield, B. and O'Mahony, M. 2004. Transit Capacity and Quality of Service Manual Applied to a Bus Corridor in Dublin, Ireland. *Transportation Research Record*, 1887, pp. 195-204.

Chu, K.K.A., Chapleau R. and Trépanier, M. 2009. Driver-assisted bus interview (DABI): Passive transit travel survey using smart card automatic fare collection system and its applications. *Transportation Research Record*, 2105, Washington, D.C., pp. 1-10.

Chu, K.K. and Chapleau, R. 2008. Enriching Archived Smart Card Transaction Data for Transit Demand Modeling. Presented at the 87th TRB annual conference, Washington, DC. *Accepted for publication in the Transportation Research Record: Journal of the Transportation Research Board.*

Conklin, J., Englisher, L. and Shammout, K. 2004. Transit Customer Response to Intelligent Transportation System Technologies Survey of Northern Virginia Transit Riders. *Transportation Research Record*, 1887, pp. 172-182

Fielding, G.J., Gauthier, R.E. and Lave, C.A. 1978. Performance indicators for transit management. *Transportation*, no. 7, pp. 365-379.

Furth, P.G., Hemily, B., Muller, T.H.J. and Strathman, J.G. 2006. *TCRP Report 113: Using Archived AVL-APC Data to Improve Transit Performance and Management*, TRB, Washington DC.

Gillen, D., Chang, E. and Johnson, D. 2001. Productivity Benefits and Cost Efficiencies from Intelligent Transportation System Applications to Public Transit. Evaluation of Advanced Vehicle Location. *Transportation Research Record*, 1747, pp. 89-96.

He, L. and Trépanier, M. 2015. Estimating the Destination of Unlinked Trips in Public Transportation Smart Card Fare Collection Systems. *Transportation Research Record: Journal of the Transportation Research Board*, 2015 (In Press).

Kusakabe, T. and Asakura, Y. 2014. Behavioural data mining of transit smart card data: A data fusion approach. *Transportation Research Part C: Emerging Technologies*, 46, pp. 179-191.

Kittelson and Associates, Inc. 1999. Transit Capacity and Quality of Service Manual (1st ed.). *TCRP Project A-15*. TRB, National Research Council, Washington, D.C.

Morency, C., Trépanier, M. and Agard, B. 2007. Measuring transit use variability with smart card data. *Transport Policy*, Vol. 14, no. 3, pp. 193-203.

Munizaga, M.A. and Palma, C. 2012. Estimation of a disaggregate multimodal public transport Origin–Destination matrix from passive smart card data from Santiago, Chile. *Transportation Research Part C: Emerging Technologies*, Volume 24, October 2012, pp. 9-18.

Nurul Hassan, M., Hawas, Y.E. and Ahmed, K. 2013. A multidimensional framework for evaluating the transit service performance. *Transportation Research Part A: Policy and Practice*, Volume 50, April 2013, pp. 47-61.

Pelletier M.-P., Trépanier, M. and Morency, C. 2011. Smart card data in public transit: A review. *Transportation Research C: Emerging Technologies*, 19(4), pp. 557-568.

Perk, V.A. and Foreman, C. 2003. Florida Metropolitan Planning Organization Reports on Transit Capacity and Quality of Service. First-Year Evaluation. *Transportation Research Record*, 1841, pp. 128-134.

Pourmonet, H., Bassetto, S. and Trépanier, M. 2015. Vers la maîtrise de l'évasion tarifaire dans un réseau de transport collectif. *11th International conference of Industrial Engineering*, Québec, October, pp. 26-28.

Raveau, S., Munoz, J.C. and de Grange, L. 2011. A topological route choice model for metro. *Transportation Research Part A*, 45, pp. 138-147.

Shelfer, M. and Procaccino, J.D. 2002. Smart card evolution. *Communications of the ACM*, July 2002/Vol. 45, no. 7, pp. 83-88.

Si, B., Fu, L., Liu, J. and Shiravi, S. 2014. A multi-class traffic assignment model for predicting transit passenger flows: A case study of Beijing subway network. *In Transportation Research Board 93rd Annual Meeting*, Washington, DC.

Sun, Y. and Schonfeld, P.M. 2014. Schedule-based route choice estimation with automatic fare collection data for rail transit passengers. *In Transportation Research Board 93rd Annual Meeting* (No. 14-0834).

Trépanier, M., Chapleau, R. and Tranchant, N. 2007. Individual trip destination estimation in transit smart card automated fare collection system. *Journal of Intelligent Transportation Systems: Technology, Planning and Operations*, Taylor & Francis, Vol. 11(1), pp. 1-14.

Trépanier, M., Morency, C. and Agard, B. 2009. Calculation of transit performance measures using smart card data. *Journal of Public Transportation*, 12(1), pp. 79-96.

Trépanier, M., Barj, S., Dufour, C. and Poilpré, R. 2004. Examen des potentialités d'analyse des données d'un système de paiement par carte à puce en transport urbain, Congrès annuel 2004 de l'Association des transports du Canada (Québec), pp. 10-14.
Trépanier, M. and Chapleau, R. 2006. Destination Estimation from Public Transport Smart card Data. *12th IFAC Symposium on Information Control Problems in Manufacturing* (INCOM), Saint-Étienne, France.
Trépanier, M. and Vassivière, F. 2008. Democratized Smart card Data for Transit Operators. *15th World Congress on Intelligent Transport Systems*, New York, États-Unis, pp. 16-20 Novembre.

## AUTHOR BIOGRAPHY

**Martin Trépanier** is Professor in Industrial Engineering at Polytechnique Montreal and co-director of the Interuniversitary Research Centre on Entreprise Network, Logistics and Transportation (CIRRELT).

**Catherine Morency** is Professor in Civil Engineering at Polytechnique Montreal and head on the Mobilité Research Chair on the Sustainability in Transportation. She is a member of the CIRRELT.

They are analysing smart card data since 2003, partnering with many transit agencies of Montreal and Ottawa, Canada. They conducted research on many aspects related to smart card data: error detection, destination estimation, KPI assessment, data mining, loyalty, influence of weather, comparison with household surveys, etc.

Chapter 11

# Ridership Evaluation and Prediction in Public Transport by Processing Smart Card Data: A Dutch Approach and Example

**N. van Oort[1,*], T. Brands[2], E. de Romph[3] and M. Yap[4]**

**ABSTRACT**

This chapter deals with the Dutch smart card system, the so-called OV-Chipkaart and illustrates a potential application of its data. This chapter explores options for using this anonymous smart card data for evaluation, analysis and performing simple what-if analyses by using transport planning software. The objective is to process the data in such a way that it enhances evaluation and prediction of ridership (patterns). This helps to improve network and time-table design. The main contribution of this research is to introduce smart card as a data source into existing methods to come to a new ridership prediction approach. Our approach takes comfort into account, since it is a relevant quality indicator, which is often neglected. We show that the effect of a frequency increase in a congested public transport line in terms of additional passengers becomes significantly larger when comfort effects are included. Our approach was applied as a case study to the tram network in The Hague. The approach proved very valuable to gain insights on the effect of changes in the public transport supply.

## 1. INTRODUCTION

Both the amount and number of sources of data is rapidly increasing in our society and thus also in the public transport industry. Almost all buses, trams and metros in the world are equipped with on-board computers and transmit terabytes of data with regard to for instance trip times, delays

[1] Delft University of Technology/Goudappel Coffeng, PO Box 5048, 2600 GA Delft. Email: N.vanOort@TUDelft.nl
[2] University of Twente/Goudappel Coffeng.
[3] Delft University of Technology/TNO.
[4] Delft University of Technology/Goudappel Coffeng.
* Corresponding author

and dwell times. Imagine social media data, such as user data of Twitter, Facebook and Flickr, which may yield new insights on public transport usage (Bregman 2012). Furthermore, video cameras (e.g. surveillance systems in stations and on-board vehicles), Wi-Fi and Bluetooth trackers provide information of pedestrian flows in stations, at platforms and in-vehicles (Van den Heuvel et al. 2015). Sensors connected to different types of assets, signals and switches for instances, enable optimization of maintenance schemes.

This chapter focuses on a potential application of smart card data. Recently, many cities and regions introduced a smart card system for their public transport systems as discussed in this book and various publications such as Pelletier et al. (2011), Ma et al. (2013), Kurauchi et al. (2014), Wang et al. (2011) and Park et al. (2008). In addition to ticket handling, being an alternative for individual regional or urban tickets, these systems also provide valuable data. Without these systems, detailed information of origin and destination, number of passengers, trip lengths, etc. can only be collected by time consuming and expensive surveys. That is why the current surveys provide limited data sets. Smart card systems have the potential of providing more and better insights of passenger behaviour. These insights are helpful when dealing with the main challenges in the public transport industry.

Within the public transport industry, we see several challenges. Due to the increased focus on cost savings, more attention to measures that increase cost efficiency of public transport is being paid. Meanwhile, passengers require higher quality of services. Although both developments seem to contradict each other, measures that serve both objectives do exist. Improving operational speed and service reliability, for instance, will lead to higher quality and lower costs at the same time, as shown by Van Oort et al. (2015b). However, to find and optimize cost-effective measures, detailed data on performance and ridership. Fortunately, the amount of data is increasing rapidly. Automated Vehicle Location (AVL) data has already been available for a long time (Furth et al. 2006, Hickman 2004) and recently much more passenger data (Automated Passenger Counting (APC) data) has become available as well (Pelletier et al. 2011). These data support public transport design and decision making, since they enable planners to illustrate the costs and benefits of certain problems and their solutions, for instance the added value of holding (Cats et al. 2012 and Van Oort et al. 2012) or optimized synchronization between tram and train (Lee et al. 2014). These costs and benefits are relevant for decision making and may be incorporated in cost-benefit analyses.

This chapter deals with the Dutch smart card system, the so-called OV-Chipkaart and illustrates a potential application of its data. Our objective is to process the data in such a way that it enhances evaluation and prediction of ridership (patterns). This helps to improve network and timetable design. The main contribution of this research is to introduce smart card as a data source into existing methods to come to a new ridership prediction approach. The outline of this chapter is as follows. Section 2 will elaborate on smart card systems in general and the Dutch smart card data specifically. Section 3 introduces our methodology to predict ridership. Section 4 is a case study, which reviews the applicability of our approach to data to predict future ridership. The conclusion and reflection on the approach are provided in Section 5. The methodology and case studies in this chapter are partly based on Van Oort et al. (2015a).

## 2. SMART CARDS AND DATA

### 2.1 Smart Card Data Applications

For analysing, designing and optimization of public transport, actual and future demand is essential. The number of passengers and passenger kilometers in the network, per line and per stop are crucial. In addition to traditional counting, smart cards can be a a rich data source for this (see also Chapter 10). In recent years, smart card data that can distinguish between service, time of day and user groups has become available. The major advantages of smart card data for transport service providers are according to Bagchi and White (2005):

- large volumes of detailed, personal travel data;
- being able to link those data to the individual card and/or traveller, enabling to identify travel patterns over longer time periods;
- having access to continuous trip data covering longer periods of time, to identify trends (for example due to season of the year) or to see the influence of incidents or engineering works;
- knowing who the most frequent customers are.

Insights that can be gained depend on the exact characteristics of the system. The number of cities or regions where smart cards are applied is increasing rapidly (see Chapter 1). Prominent examples are London (Oyster card) and Hong Kong (Octopus card), but many more examples can be found in literature, (e.g. Seoul (Park et al. 2008), Beijing (Ma et al. 2013), Santiago de Chili (Munizaga and Palma 2012), Shenzen (Hasan et al. 2013) and Brisbane (Neema et al. 2015). Depending on the technology used, limitations in applications arise. For example, on London buses passengers need to only tap in and the destination stop needs to be estimated (Wang et al. 2011). In Ma et al. (2013), an example of Beijing is presented where no location information is connected to the smart card data and several data sources are connected to obtain the information. Another example from Quebec, Canada is Morency et al. 2007, where a method to estimate the alighting locations is given, since the smart card data does not provide this. Bagchi and White (2005) state that travel purpose is hard to obtain from the data, but in some cases this may be estimated based on the used fare type (for example a student's reduction fare) or other methodologies discussed in this book.

Canadian researchers (Pelletier et al. 2011) present a broad overview of applications of smart card data, varying from strategic and tactical planning optimization to operational improvements. Most applications aim to assess OD-patterns (Munizaga and Palma 2012 and Trepanier et al. 2007), route choice behaviour (Schmöcker et al. 2013) and transfer analysis (Seaborn et al. 2009). Surprisingly, improved forecasting based on historical data is only mentioned once.

## 2.2 The Dutch Smart Card System: OV-Chipkaart

In recent years the Dutch smart card, the OV-Chipkaart, has been introduced (Cheung 2006). This system replaced the former payment systems for regional public transport called Strippenkaart and paper train tickets. Strippenkaart was introduced in 1980 as one nation-wide payment system to replace all individual urban and regional systems. It could be used in the entire country. The fare depended on the number of zones through which one travelled. The size of these zones differed per region and so did the total price. The advantage of this system was that everybody could travel with one ticket in buses, trams and the metro throughout the country. However, for the national train services, a separate ticketing system existed. For the regional public transport operators the major disadvantage was that no information was available on where people travelled. The location of where the tickets were sold (shops and counters) was known, but not where they were actually used. Expensive surveys were required to determine how the total revenues should be split over the operators. To solve this, the public transport operators started to develop a smart card system in 2001. The system was introduced in Rotterdam in 2005 and in the rest of the country by 2012. In 2014 the last paper train tickets in the national train system were abolished.

The Dutch smart card is a nationwide system for all public transport in The Netherlands (bus, tram, metro and train). The card is used to pay the fare and on many lines (including the train) it is the only valid ticket. The system uses nfc-chip technology and passengers have to check in and to check out. Therefore, valuable information is measured about both origins and destinations of all public transport users (on station/stop level). In The Netherlands, the check in and check out devices are either located on the platform (for trains and metros) or inside the vehicle (for buses and trams). The most detailed information is available in the latter case, where each trip in a journey is tracked (a journey may consist of multiple trips, with a transfer in between). The route through the public transport network is therefore completely traceable. In case the smart card devices are located on the platforms, only information is available of the first and the last station, making route search through the public transport network necessary for analysts. In the rest of this chapter we describe the situation where information on the check in stop and the check out stop is available. This is the case in a majority of the regional and urban public transport lines in The Netherlands (all bus and tram lines).

In 2014, 10 million smart cards were in active use in The Netherlands. Every week about 2.8 million people travel using their smart card to travel in The Netherlands, producing about 42 million transactions per week. These transactions are mainly check in and check out transactions, but also include topping up the balance on the smart cards.

## 2.3 Dutch Smart Card Data

An example of the raw data format resulting from the Dutch smart card transactions is given in Table 1. Every record contains a trip, with a check in station, check in time, check out station and check out time. The anonymous smart card ID supports the combination of multiple trips into one passenger journey, by identifying transfers that are made. Furthermore, a public transport line number is given, so that the trip may be matched to a specific service in case multiple public transport lines run parallel. Potentially, the vehicle number and/or run number are also given. In that case, detailed analysis of distribution among individual services is enabled, typically to provide solutions for capacity problems. Furthermore, some information may be provided on smart card type/fare type to predict trip purpose. For example, an annual ticket is usually used for commuting to and from work, while a student card is usually used for visiting schools or universities. Special offer tickets are most of times used for recreational purposes.

**Table 1.** A sample of fictitious smart card data: every record represents a trip in a public transport vehicle

| Chip ID | Check In Stop | Check Out Stop | Check In Time | Check Out Time | Line Number | (Vehicle Number) | (Ticket Type) |
|---|---|---|---|---|---|---|---|
| 1 | 35 | 488 | 10:27 | 10:52 | 9 | .. | Regular single |
| 2 | 23 | 86 | 8:01 | 8:09 | 1 | .. | Student |
| 2 | 86 | 90 | 8:17 | 8:55 | 3 | .. | Student |
| 3 | 73 | 94 | 7:20 | 7:53 | 4 | .. | Annual ticket |
| 3 | 94 | 73 | 16:55 | 17:27 | 4 | .. | Annual ticket |

For simplicity, all data in this example is for a specific date.

In Table 1, the first trip is the only trip conducted on this day by chip ID 1. This may be a trip for visiting family (including an overnight stay) or the return trip might be made by car (probably as a passenger). The second and third records are from the same chip ID. Furthermore, the trips are very close to each other in time, so we can assume these trips are a part of the same journey, which includes a transfer. We observe that the alighting stop for the first trip and boarding stop for the second trip are the same. Note that this is not necessary to form a transfer: there may be a short walking leg may in between, for example, in a large station with various tracks where there are several on-street stops. Finally, the last two records are of the same chip ID as well, but these trips are apart from each other in time, so these trips cannot be one journey. We observe that this is a typical commuting pattern: in the morning the traveller goes to work and returns home in the evening. The ticket type 'annual ticket' is another indication of commuting to work.

The technical system of the Dutch smart card system that generates the above mentioned data contains several components; see Figure 1,

- *Level 0*: The smart card. These cards always have an ID number. Both personal cards and anonymous cards exist. Personal cards can be used to load personal products, for example to get discounts or unlimited travel.
- *Level 1*: Devices that have direct contact with the smart cards: check in and check out devices and ticket vending machines. These devices have the power to change the balance on the smart card. They also provide the smart card with a check in tag, as a proof of a valid ticket in case of ticket inspection.
- *Level 2*: Local systems at public transport operators that collect data of individual transactions from level 1 devices and temporarily store it (for example located at a bus garage).
- *Level 3*: Central system for each public transport company, where all the data of a company is available and prepared to send to the national (public transport smart card) data collecting agency (i.e. TLS). At this level data of check ins and check outs to trips and the data might be added to the trip, like distance travelled or fare paid.
- *Level 4*: The database of the national data collecting agency. Here smart card transactions are verified and the financial consequences of the transactions are determined (all payment are registered). Another function is to provide personal transaction history to smart card users.

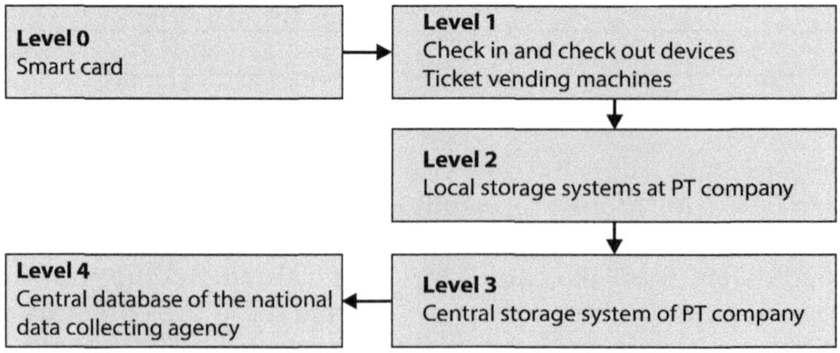


**Fig. 1.** Dataflow through the Dutch smart card system from the smart card in level 0 to the national database in level 4. A distinction may be made between components of the national agency (left) and components of the public transport companies (right)

Technically speaking, the data is available at the individual level, giving large possibilities for detailed analysis. However, there are some concerns about the availability of the data and privacy agreements that must be taken into account. Privacy is the most important issue, as individual data is used. Therefore, Dutch privacy law states that processing individual data is not allowed and that data must not be preserved for more than 18 months. It is also required that before the start of any research in which smart card data is used, the objective should be clearly stated. The dataset cannot be used for other purposes.

Another concern is the availability for analysis. The data is owned by public transport operators and most of them see it as confidential company information, due to the tendering system of the Dutch public transport concessions. Data of only one public transport operator is available for analysis, since this could be regulated in contracts. However, combining data from more operators is still difficult due to this issue. This data could be valuable, for example as to analyse movements in a train station. Currently both the national and regional governments are trying to solve these issues with the co-operation of the operators. The first attempt of connecting data of operators is described in Nijenstein and Bussink (2015). They show how trip chains of smart card data of multiple operators [HTM (urban public transport operator in the The Hague region) and NS (national railways)], was created.

## 3. PREDICTING RIDERSHIP BY SMART CARD DATA

### 3.1 Introduction

Making predictions for public transport can be done in several ways, ranging from multimodal activity based models to simple rules using spread sheets. In The Netherlands, a hierarchy of traffic forecast models exists. The national model is a disaggregated model mainly focused on road travel. Four more detailed regional models exist using the same principles as the national model, but with more detailed networks. Public transport is modelled during the distribution phase of the model, but the level of service matrices are mainly exogenous (Joksimovic and Van Grol 2012).

At the urban level, many cities in The Netherlands have their own models. In most of these models public transport is modeled in more detail on the network level. However, the models are generally simpler. Most of them are multimodal gravity models for estimating the demand. Recently, the importance of the bicycle as access mode to public transport was recognised, resulting in more sophisticated models using a nested logit structure that distinguishes between different access and egress modes in public transport (Brands et al. 2014).

Most of the public transport operators in The Netherlands do not use transport models for predicting ridership or changes in demand. For shorter term changes in their services, due to maintenance mostly spread sheets are used with relatively simple rules. Transport models could provide valuable insights for public transport operators. Many regions, however, do not have a multimodal transport model, or the level of detail of these models does not match the level of operation within the public transport company.

With the introduction of the smart card system several public transport operators wonder what they can do with this massive data. Mainly because of the continuous nature of this data, systems and ideas have emerged to use this for gaining more insight into current use. There are numerous studies wherein smart card data is analysed specifically for the current situation in order to replace survey data. They mainly focus on aspects as data cleaning, estimating alighting stop, etc. For an overview of these studies, see Pelletier et al. (2011). There are several studies wherein origin destination information (OD-matrices) is derived from the data, either in aggregated form (Wang et al. 2011) or disaggregated form (Munizaga and Palma 2012 and Bouman et al. 2013).

Once a matrix is produced that reproduces the passengers in the services, it becomes possible to perform what-if analysis, by assigning this matrix to the public transport network. The simplest possibility is to assume the demand remains fixed but this would show only the effect on route choice. A better approach would be to assume the demand matrix reacts to changes in the network. This could be done using a simple elasticity model. When a full multi-modal model exists for the study-area, the demand shift could be taken from the model. The most feasible approach depends on the availability of existing models and the time horizon for decision making. Table 2 gives an overview of the possibilities.

**Table 2.** Possible public transport models

| | Multimodal Model | Elasticity Model | Quick-Scan Model |
|---|---|---|---|
| **Modes** | Car, public transport, bike | Public transport | Public transport |
| **Scale** | National, regional, urban | Regional, Urban | Urban |
| **Time Horizon** | 10-20 years | < 10 years | < 5 years |
| **Project Type** | Strategic, policies, infrastructure changes | Tactical, changing lines, frequencies, stops | Tactical, changing lines, frequencies |
| **Usage** | Modal split, cost-benefit analysis | Network effect | Route choice effects |

In the following we derive and apply an elasticity model with the smart card data. These kind of models are relatively simple to construct (thus saving time and budget) and can make good use of the available data. The accuracy level is lower than multimodal models, but still enough for several research objectives.

## 3.2 Deriving OD Demand from Smart Card Data

The derivation of OD matrices from smart card data makes it possible to perform what-if analysis using traditional transport modelling. The smart card data produced by the Dutch system does not have many of the problems seen in other studies due to check in and check out data being available. This makes the construction of trips made by individuals from first boarding to last alighting including transfers between services possible. In urban public transport in The Netherlands, smart card data transactions take place in each vehicle separately. This means that a traveller has to check in and check out in each vehicle. If a transfer is made, two separate transactions are registered. To estimate an OD-matrix, it is necessary to aggregate these trips made by one traveller to an origin-destination level. Three aspects are especially relevant in this process, are highlighted in the following subsections.

### *3.2.1 Threshold Time for Transfers*

First, it is important to determine a valid threshold time to combine two trips with a transfer in between to one total trip. If a traveller spends a longer time period at a station than this threshold, the trip is seen as a new trip. Using a very short threshold time between two subsequent trips may wrongly interpret trips with a relatively long transfer time. This is especially relevant when transfering to a public transport service with a low-frequency, where waiting times can be long. In this situation with a very strict threshold – for example a threshold value of 10 minutes – some trips with an intermediate transfer might not be correctly aggregated to the OD-level. This may lead to overestimation of the number of trips (by considering a trip from A, via B, to C, as two separate trips A-B and B-C), to an underestimation of the trip length, underestimation of the number of transfers and to biased errors in demand on OD-pairs. On the other hand, when applying a very high threshold value – for example 60 minutes – probabilities increase that two separate trips back-and forth are ignored. For example, consider a traveller leaving from origin A to destination B, performing a short activity and then returning from origin B back to destination A. In this situation with a high threshold value, these two separate trips will be wrongly aggregated to one trip with the same origin and destination, namely A. Usually trips with the same origin and destination are excluded from analysis, this leads to underestimation of PT ridership. It is therefore important to find a balance between these issues by applying an intermediate threshold criterion to decide whether to aggregate two separate trips to one OD-trip. In The Netherlands, 35 minutes is often used, because this is the threshold value used for the fare system. This fare system is distance based, but also includes a fixed start fare, which has to be repaid if a traveller boards the next vehicle more than 35 minutes after leaving the previous vehicle. In urban areas, where the frequencies are usually more than twice per hour, a somewhat lower value (i.e. 25 minutes) seems more appropriate, given the occurrence of the effect of high threshold values described above.

### 3.2.2 Unique Card Number

Second, one should be aware that a unique card number is required to determine a transfer made by a certain traveller to aggregate two trips with an intermediate transfer to one OD-relation. Due to privacy regulations, it can be difficult to get transaction data from public transport operators or authorities with a unique card number for each transaction, but after aggregation of trips to OD pairs, this card number is not needed any more and can be deleted. If needed, this aggregation may already be done by the public transport operators themselves. It is important to use a relatively large dataset to determine an OD-matrix. For example, using smart card transactions from 20 workings days as input to estimate an OD-matrix for an average working day increases the probability that trips are made between all relevant OD-pairs. This implies that non-integer numbers may occur in the OD matrix, to represent demand on OD-pairs with only occasional demand. On the network level, the demand will be as realistic as possible in this way. Next to this, one should be aware that some travellers use different smart cards for different parts of their trip. For example, someone might use his/her private smart card for the travelling from home to the train station and use a company-owned smart card for the main part of the journey. Because these cards have different card numbers, it is not possible to aggregate these trips to the OD-level. This means that this trip will be reflected in the OD-matrix as two separate OD-trips.

### 3.2.3 Time Dependent OD Demand

Detailed information is available in the smart card data about check in and check out time. This enables the modeller to define any desired time period. In the current application, the average between first-boarding and last-alighting time is taken to determine in which hour of the day the trip took place.

In the next step, these hours of days are aggregated into typical modeling periods, like AM peak, PM peak, day time and evening time. For a different application, typically when capacity is relevant, more detailed time periods are possible, for example half hour periods. This may reveal that for example between 7:00 and 7:30 the number of passengers is still limited, while from 7:30 the system is packed with passengers. This cannot be revealed on the hour level, neither on the AM peak level. On the other hand, smaller time periods also have difficulties, because many trips are longer than 15 minutes. This makes the assignment of a trip to one time period problematic, because many trips occur in various time periods.

A matrix between stations should ideally be converted to a zone-to-zone matrix. In this study this step is omitted. The resulting station-to-station matrix should be assigned to the network in order to check the measured number of passengers in the services is reproduced. This requires calibrating the assignment model's route choice parameters which might also be done with smart card data (see Chapter 4). In this study this was done manually.

### 3.3 Elasticity Model

Given an OD matrix that is provided by smart card data, a public transport network and a calibrated route choice and assignment model, the step towards short and medium time prediction can be made. Such a tool would allow one to assess the network effects of changing the frequency of lines, changing routes of lines, introducing new routes and increasing the speed of a line. These measures may be temporary or permanent. Next to changes in route choice, changes in demand could also be expected.

In this chapter we present a method that is based on demand elasticity: the relative change in costs per OD pair have an effect on transportation demand on that OD pair. For a good overview of elastic demand models (see Litman 2013).

The costs of a public transport trip comprise of several components: in-vehicle time, waiting time, number of transfers (penalties) and fare. All these components of the trip are expressed in monetary values by the coefficients and summarized. The resulting value is referred to as the generalized cost between stop $i$ and stop $j$: $C_{ij}$. For the Dutch situation the values for $\alpha$ are known and taken from literature (Significance et al. 2013 and Wardman 2004).

For in-vehicle time 6 Euros per hour is used in Significance et al. 2013. For waiting time, a factor is used that is one and a half times as high as the factor for in-vehicle time (i.e. 9 Euros per hour) (Wardman 2004). For transfers a penalty of 5 minutes is used, which means a cost of 5 times 9/60 Euro for every transfer.

Equation (1) shows the calculation of generalized costs for OD pair $i,j$. The coefficient of fare $\alpha_4$ is equal to 1, because it expresses the costs in monetary values.

$$C_{ij} = \alpha_1 T_{ij} + \alpha_2 WT_{ij} + \alpha_3 NT_{ij} + \alpha_4 F_{ij} \qquad (1)$$

*With*:

| | |
|---|---|
| $C_{ij}$ | Generalized costs on OD pair $i,j$ |
| $\alpha_1,\alpha_2,\alpha_3,\alpha_4$ | Weight coefficients in generalized costs calculation |
| $T_{ij}$ | In-vehicle travel time on OD pair $i,j$ |
| $WT_{ij}$ | Waiting time on OD pair $i,j$ |
| $NT_{ij}$ | Number of transfers on OD pair $i,j$ |
| $F_{ij}$ | Fare to be paid by the traveller on OD pair $i,j$ |

Figure 2 shows the steps in our elastic demand calculation. First, using a public transport route choice algorithm (Brands et al. 2014), the generalized cost matrices are calculated for the base situation and the situation that includes a network scenario. Note that this requires successful calibration of the route choice parameters: we here assume that the route choice algorithm is able to reproduce the line loads in the base situation. Comparing the cost matrices results in relative cost changes per OD pair. Using the OD matrix for the base situation (from smart card data) and an elasticity value (Wardman 2012, TRB 2004, Balcombe et al. 2004), the relative changes in OD flows are calculated, resulting in an OD matrix for the network scenario. Importantly, the availability of smart card data offers great opportunities to assess new elasticity values by performing revealed preference research. New values may be found for both structural and temporal changes in offered service quality (see Van Oort et al. 2016). The final step is to assign this OD demand to the public transport network, again using the public transport route choice algorithm.

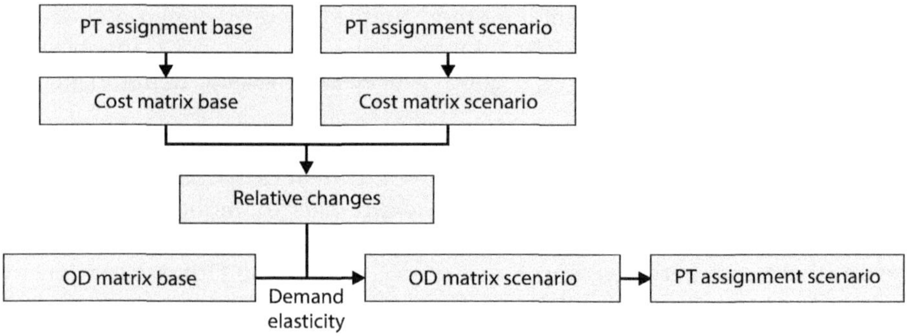


**Fig. 2.** Schematic representation of the demand prediction model

The elasticity model used in this study is captured in Equation 2. The new OD demand is calculated (in the situation with the network scenario) from the base demand using the change in cost and the elasticity value. The subtraction and later addition of 1 in the equation is to convert from a growth factor to relative growth or vice versa. Accordingly, in this definition the value for elasticity should be negative to be realistic, since an increase in costs then leads to a decrease in demand. Consequently, the demand change is directly calculated from generalized costs. This is different from using, for example, travel time elasticity or fare elasticity, since those values only include specific components of the generalized costs. The value of generalized costs elasticity is chosen in such a way that the effect of a travel time or fare change roughly corresponds with the changes that would occur when using travel time or fare elasticity.

$$D_{ij}^{1} = \left( E\left( \frac{C_{ij}^{1}}{C_{ij}^{0}} - 1 \right) + 1 \right) * D_{ij}^{0} \qquad (2)$$

*With*:

$D^1_{ij}$ Demand on OD pair *i,j* in the scenario

E Elasticity

$C^1_{ij}$ Generalized costs in the scenario

$C^0_{ij}$ Generalized costs in the base situation

$D^0_{ij}$ Demand on OD pair *i,j* in the base situation

Extensions of this model can be made when new housing or job developments take place in the region at study. The relative growth of housing or jobs around public transport stops may be converted into growth factors to be applied to rows or columns of the OD matrix. Then the assumption is made that the distribution of trips among destinations or origins does not change from the observed distribution (based on smart card data) in the base situation. When both rows and columns are adjusted, a balancing method should be applied, for example the Furness method.

## 3.4 Incorporating Comfort Impacts

### *3.4.1 Effects of Comfort and Crowding*

In urban public transport systems, there are increasingly problems regarding the supplied comfort and capacity to passengers. Comfort is however hardly incorporated in current public transport demand models, although comfort and crowding levels influence passengers' route and mode choice. Besides, there is no unlimited capacity available on public transport lines. For a realistic modelling of passenger demand and route choice in crowded public transport systems, it is important that the number of passengers assigned to a public transport service does not exceed the capacity of this service. Therefore, as an extension of the elasticity demand model described in the previous section, comfort effects are incorporated in the predictions. Smart card data also proved to be very supportive in this process. Such a model extension is useful when a heavily used public transport network is studied. This can be in a temporary context, for example when due to engineering works a public transport line is not available anymore, and the public transport operator wants to check whether the capacity of public transport lines on alternative routes is reached, in order to anticipate on this accordingly. Also in a permanent situation this may be relevant: when the frequency on a crowded high frequency line is increased, the current models only predict a limited passenger growth due to slightly reduced waiting times. However, in reality passengers may decide to start using the service due to an increased comfort level as well.

In literature several approaches exist to assign traffic to a public transport network, incorporating capacity. Cepeda et al. (2006) put a hard capacity constraint to public transport links, while Florian (2002) uses a crowding function where link costs depend on the flow. Pel et al. (2014) go one step further by introducing a crowding function both upon boarding (using an additive trip penalty) and in public transport line sections (using a time multiplier). Schmöcker et al. (2011) use a similar approach by introducing the "fail-to-sit" probability. This requires a Markov type network definition with two states: sit and stand, using priority rules when passengers change state. Pel et al. (2014), Schmöcker et al. (2011) and Cepeda et al. (2006) apply their methods to real world case studies, but only as an assignment method in itself (not as a part of a larger modelling framework).

In general, crowding in public transportation can have three different effects:

- The in-vehicle time perception of travellers increases with a more crowded vehicle, since a crowded vehicle is perceived as less attractive than a quiet vehicle.
- Passenger demand for a certain PT service exceeds the supplied capacity. On the short run this denied boarding leads to passengers having to wait another interval time for the next vehicle. On the longer run, for permanent and published maintenance works, an equilibrium situation will rise where passengers adjust route and mode choice such that supplied capacity is not exceeded. For unplanned disturbances, no equilibrium situation is expected.
- Dwell time of public transport vehicles increases with higher crowding levels, since the boarding and alighting process will take more time.

In this study, we focus on incorporating the first two comfort effects in PT demand models.

### 3.4.2 Methodology of Incorporating Comfort

The two mentioned comfort effects are incorporated in our model by making the in-vehicle travel time component of the generalized costs function dependent on the passenger load. For this, a crowding function is used. The perceived in-vehicle travel time is calculated as a multiplication factor over the real, objective travel time, which depends on the passenger load in relation to the number of seats and to the capacity for standing passengers. First, the transformed volume/capacity (VC) ratio is determined using Equation 3. The result of this formula is that VC = 1 when the passenger load $L$ equals the seat capacity $C_{seated}$ of a certain vehicle. VC = 2 when the load $L$ is equal to the crush capacity (seated plus standing passengers) $C_{crush}$. The seat capacity and crush capacity can be specified for each public transport line and each modelling period (morning peak, evening peak, off-peak hours) separately, in order to distinguish between different vehicle types and lengths used on different lines during different times of the day.

$$VC = \begin{cases} \dfrac{\dfrac{L}{C_{seated}}}{1+\dfrac{L-C_{seated}}{C_{crush}-C_{seated}}} \end{cases} \tag{3}$$

Most studies on valuation of crowding only use the load factor – the passenger load $L$ divided by the seat capacity $C_{seated}$ – to express crowding effects (Wardman and Whelan 2011). In our study, we explicitly distinguish between the seat capacity $C_{seated}$ and the crush capacity $C_{crush}$ of public transport vehicles. Taking both the seat capacity and crush capacity into account has the advantage that different types of vehicles with different configurations (with relatively less or more seats with respect to the total capacity) become comparable. In a public transport vehicle with a relatively high number of seats relative to the total crush capacity (e.g. an intercity train service), crowding will be perceived differently compared to a vehicle with a relatively low number of seats in relation to the crush capacity (e.g. a light rail or metro service). This means that in reality the load factor only makes sense, when it is related to the total crush capacity of a vehicle.

In their meta study to crowding valuation in public transport, Wardman and Whelan (2011) indicate that the in-vehicle time multiplier should be expressed as function of the load factor, up to a load factor of 100% of the seat capacity $C_{seated}$. For highload factors, the vehicle configuration needs to be considered as well. For load factors between $C_{seated}$ and $C_{crush}$ we determine the in-vehicle time multiplier as function of both the seated and crush capacity.

Based on the VC ratio, a piecewise linear function is used to determine the factor for perceived travel time F, based on the values in Table 1. Starting from 80% seat occupation the comfort level starts to decline following Douglas Economics (2006). According to Douglas Economics, the multiplication factor equals 1.1 when a 100% seat occupation rate is reached. Revealed occupation rates using smartcard data are used to determine the crush capacity of different types of public transport vehicles. The crush capacity as specified by the manufacturer, assuming 4.5 persons/m2, appears not to be realized in practice in the Netherlands. Based on vehicle configuration and the maximum number of passengers per vehicle found in actual smartcard data of tram lines in The Hague, we determined that the crush capacity $C_{crush}$ in the vehicles in our study is reached with 3.5 persons/$m^2$. Using the crowding multipliers from MVA Consultancy 2008 – where seated and standing multipliers are expressed as function of the number of standing passengers per $m^2$ – we determined that the multiplication factor increases with 0.64 from $C_{seated}$ to $C_{crush}$. Wardman and Whelan (2011) conclude that the in-vehicle time perception increases linear with increasing crowding levels. Non-linearity's could not be justified empirically. This leads to a piecewise linear function with crowding multipliers as shown in Table 3.

**Table 3.** Relation between *VC* and the perceived travel time factor. A factor of 1 implies no additional perceived travel time

| VC | Perceived Travel Time Factor F |
|---|---|
| 0 – 0.8 | 1 |
| 0.8 – 1.0 | 1 – 1.1 |
| 1.0 – 2.0 | 1.1 – 1.74 |
| 2.0 – 3.0 | 1.74 – 10 |

Using Equation 4 this factor is applied over the real link travel times to calculate the perceived travel time, which replaces real travel time in the generalized costs function (Equation 1).

$$T_{ij}^{per} = T_{ij} * F \tag{4}$$

To prevent the assignment of passengers to a vehicle where $C_{crush}$ has already been reached, the VC function increases steeply for VC values > 2.0. In this way, the attractiveness of a route with a completely crowded vehicle decreases in such way, that passengers will change their route or mode choice. This leads to the crowding function as visualized in Figure 3.

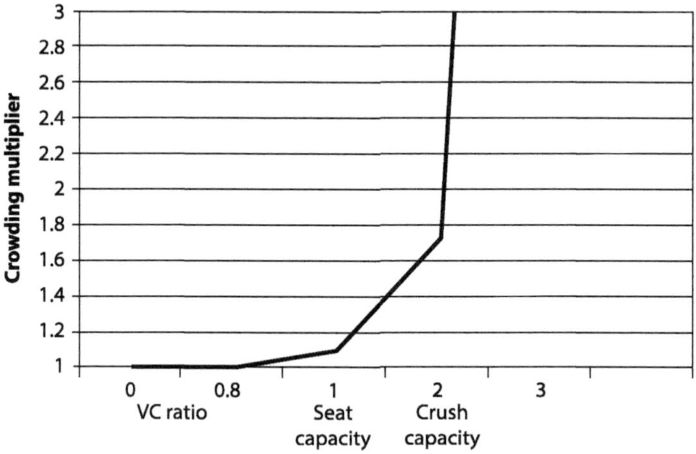


**Fig. 3.** Crowding function

Note that the load is needed for a 1 hour time period, because the capacity is also given per hour (resulting from the frequency and seat/crush capacity per vehicle). If the modelled time period is longer, a correction factor is used. Depending on the evenness of the load distribution over this time period, this factor is equal to the period length in hours (in case of a perfectly uniform distribution), or is smaller than the period length. If the distribution is uneven; the busiest hour is taken as representative for the entire time period, by dividing the real number of hours by the busiest hour factor.

Since the costs of travelling now depend on the load, an iterative assignment is necessary. This assignment procedure is comparable to a user equilibrium assignment which is common in road network assignment when incorporating congestion effects. The iterative procedure is repeated until convergence is reached between iterations $N$ and $N + 1$. We specified a convergence criterion of 5%.

# 4. CASE STUDY: THE TRAM NETWORK OF THE HAGUE

## 4.1 Introduction

We applied our approach of connecting anonymous smart card data to a transport model and performing predictions (with and without incorporation of comfort effects) in a case study. In this case study, we tested whether our approach presented in the previous section would work with actual data and real life networks. We connected anonymous smart card data of HTM, the tram operator in The Hague (about 500,000 inhabitants, 3rd largest city of the Netherlands) to a transport model built in OmniTRANS (http://www.dat.nl/en/products/omnitrans/). The city of The Hague has 12 tram/light rail lines with a total length of about 335km. Yearly, about 70 million passenger use these lines. In addition to these tram lines, the public transport in and around the city consists of urban and regional bus lines and railway lines. In this case, we have only investigated the tram lines. This made the calibration of the route choice model relatively simple because in most cases only one route choice option existed. In future research we will add the bus and train services to the model. This will give us more possibilities to calibrate the route choice model. The calculation time required for the prediction of new demand for one time period (i.e. a morning peak) including a complete iterative elastic assignment is around 25 minutes on a regular core i5 laptop.

## 4.2 Evaluation

A first step in supporting public transport planners and designers is visualizing historical (smartcard) data. In, for instance Van Oort and Van Nes (2009) and Van Oort et al. (2015c), examples of AVL data visualization are provided. In addition, illustrating smart card data on a geographical layer is beneficial as well.

To combine anonymous smart card data with geographical information, we imported the public transport network into the software environment OmniTRANS using timetable data which is publicly available in GTFS (General Transit Feed System) format. This format was introduced by Google to allow public transport operators to feed their timetables to Google Maps. This data contains the lines, positions of stops and the departure and arrival times of each run at each stop. It is translated into frequencies and travel times per line per time period (AM peak, PM peak, off-peak day period and evening). The information of the lines (including the locations of stops) is mapped geographically on the underlying infrastructure, in this case the tram rail network of The Hague. The resulting network can be seen in Figure 4.

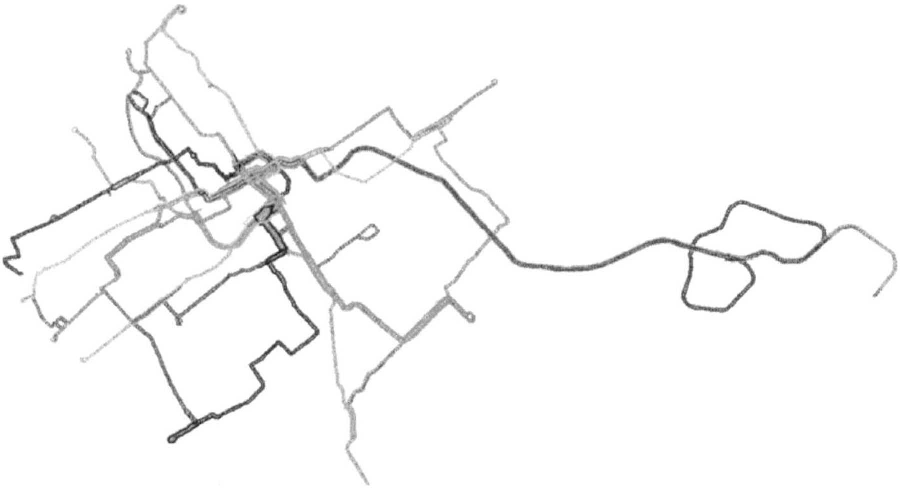

**Fig. 4.** All lines in the tram network of The Hague

In this case a decision was made to put the zones directly at the stops. The combination of geographical data of stops and lines and the smart card data (average working day of one month) is used to visualize passenger flows on the network. To this end, the smart card data (in the format of Table 1) is first pre-processed: invalid records are removed (for example records with the same stop for check in and for check out), being less than 5% of the total data amount, and trips are combined to journeys by identifying transfers, based on smart card ID and check out/check in time. After that, the journeys are loaded onto the network, following the check in and check out stop and public transport line number in the data. When the network data (from GTFS) and smart card data (from the public transport company) of the same date are used, these two data sources fit very well: almost all records from the smart card data can be directly imported.

The resulting geographical visualization can be shown over time, since the check in and check out times are known. Given the assumption that the time stamp determines the time block of the trip (check in time, check out time or an average between the two), the data can be visualized per aggregated time period, for example per one-hour period. Figure 5 shows the link loads in the AM peak for a one-day sample: it can clearly be observed that before and after the peak period, the flows are much lower than during the peak period (see the presented loads in the added circle for instance). This time-dependent data may as well be visualized in an animation. The visualization helps to understand the past: identifying high or low flows, identifying important (transfer) stops and understanding the difference among time periods over the day.

**Fig. 5.** An example of time dependent visualization of smart card data (link loads): upper left: 6 AM-7 AM; upper right: 7 AM-8 AM; lower left: 8 AM-9 AM; lower right: 9 AM-10 AM.

## 4.3 Predicting

### *4.3.1 General Case*

In addition to showing historical data, by connecting network and smart card data in the transport model, we also tested our elasticity method on the actual data simulating several measures. We investigated frequency changes, fare adjustments and rerouting of a line. The elasticities we used were based on literature (Balcombe et al. 2004) and also on rules of thumb of HTM. For instance, we used an elasticity value of -0,5 for travel time changes (E in Equation 2). This means that an increase of 10% in travel time will lead to 5% less travellers. The rules of thumb of HTM were audited and proven to be valid by independent research (Oostra 2004).

We adjusted the original skim matrix to the measures and calculated the new passenger OD-matrix accordingly. We assigned this matrix, using the Zenith-algorithm (Brands et al. 2014). Similar visualizations as shown in Figure 5 may be generated showing the new link loads. Figures 6 and 7 show the outcomes (in terms of change in passenger load) of two examples of specific network scenarios: a frequency increase and a route change in a public transport line. The main contribution of this method is that we clearly see the network impacts. Figure 6 shows a frequency increase on two lines, with expected ridership growth on these two lines (green), but also a decrease in a nearby line (red). With Figure 7 we illustrate the impacts of a route change (a link was blocked and trams had to be diverted). Due to higher travel costs on the new route, the total number of passenger decreased (increase on the diverted line route (green) is smaller than decrease on the original route (red)).

**Fig. 6.** The effect of a frequency increase on ridership

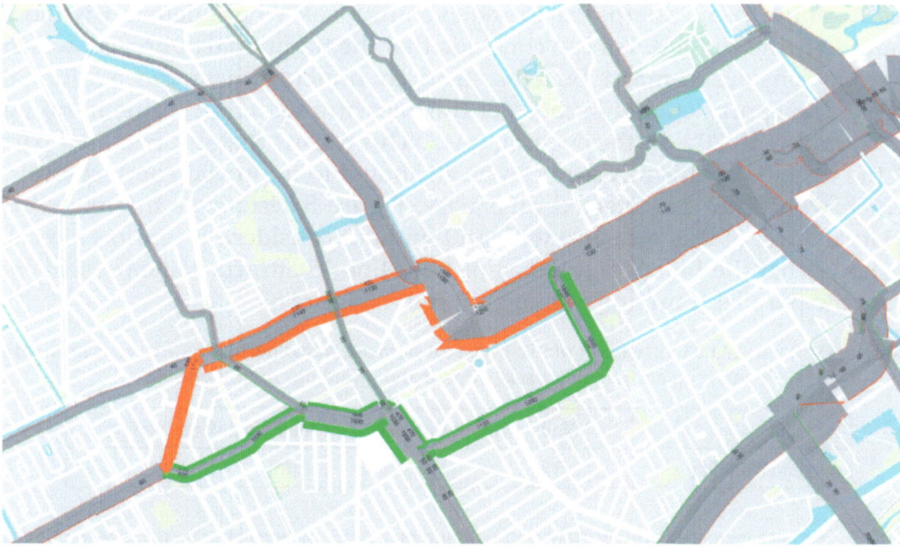

**Figure 7.** The effect of a change in route on ridership

We did a validity check on the results, which were in line with the existing methods (traditional models). However, the next step would be detailed research on revealed behaviour after changes to find updated elasticity values, specifically focusing on this area and the types of passengers.

### 4.3.2 Results after Incorporating Comfort

To see whether our approach of incorporating comfort in the prediction process is applicable and valid, we also applied this method in a case study in The Hague. The case consists of increasing the frequency of tram line 15 (line length: 9.4 km) from 6 to 8 trams per hour during the morning and evening peak. Since this tram line has a high peak demand, the effect of an increase in frequency on public transport demand is investigated for the situation with and without incorporating comfort effects of this measure.

Table 4 shows the predicted relative effect of the frequency increase on public transport demand for tram line 15 without and with considering comfort effects. Table 5 shows the expected absolute increase in public transport demand as consequence of this measure on the public transport network as a whole, considering substitution effects between lines as well. From this table we can conclude that 165 new passengers are expected in both the morning peak and evening peak, when only benefits from a reduced average waiting time are considered. When both the effects of reduced waiting time and improved comfort are incorporated, 240 and 200 new passengers are expected in the morning and evening peak respectively. Since in the morning peak public transport demand is more clustered within a small period, comfort benefits of this measure are larger during the morning peak, compared to the evening peak where demand is more uniformly distributed. We can conclude that the traditional approach, which do not consider comfort benefits, tend to underestimate the additional public transport demand because of this measure with 30% in the morning peak, and with 20% in the evening peak. This means that a substantial part of the benefits of this measure can be attributed to improved comfort levels, which would not be detected otherwise. Figure 8 visualizes the modelled relative effect of this measure with and without considering comfort effects. It shows that the higher frequency of tram line 15 attracts some passengers from the parallel tram line 1 (shown in red in Figure 8).

**Table 4.** Estimate relative increase in public transport demand tram line 15 after frequency increase in morning and evening peak (without and including comfort effects)

| | Model without Comfort | Model including Comfort |
|---|---|---|
| **Average Work Day** | +8% | +10% |

**Table 5.** Estimate increase in public transport demand on a network level after increase of frequency in morning and evening peak (without and including comfort effects)

| | Model without Comfort | Model including Comfort |
|---|---|---|
| **Morning Peak** | + 165 | + 240 |
| **Evening Peak** | +165 | + 200 |

**Fig. 8.** Relative network effects of frequency increase on link loads a) without considering comfort (left) and b) considering comfort effects (right) during morning peak

## 4.4 Reflection

In this research we have chosen an initially practical approach by choosing a zonal system corresponding to the actual stops. A more desirable option is to choose a zonal system as used in the model system for this region (resulting in OD-matrices from zone-zone instead of stop-stop). This would allow direct usage of modal split factors from this model while having a more accurate matrix for the current situation.

Because our case was limited to tram lines only, route choice in this network did not play a significant role. This meant that calibrating the route choice model using smart card data was not very challenging as mostly just one route was feasible. In an anticipated extension of this study we intend to increase the network with all the bus lines of The Hague, resulting in a network with significant route choice options. A public transport route choice algorithm needs parameters, for example logit parameters in stop choice and line choice models, or the weight factors for cost components in the generalized costs function (as are also defined in this chapter). The detailed smart card data presented in this chapter could be used to calibrate these parameters, since the actual routes chosen by travellers are observed (i.e. line choice and transfer stop choice). If multiple routes are available, a distribution among the routes can be derived from the data that should be estimated by the route choice model and its parameter settings. Furthermore, it can be tested whether these parameters are approximately equal for different situations (i.e. short trips vs. long trips).

## 5. CONCLUSIONS

Public transport operators are exposed to massive data collection from their smart card systems. In recent years The Dutch smart card, the OV-Chipkaart, has been introduced. This smart card system covers all public transport in the Netherlands (bus, tram, metro and train). The system was introduced in Rotterdam in 2005 and in 2012 the full country was equipped. In 2014 the last paper train tickets in the national train system were abolished. Every passenger needs to check in and check out, resulting in detailed information on the demand pattern. In buses and trams, check in and check out take place in the vehicle, providing also information on route choice.

This chapter explored options for using this smart card data for evaluation, analysis and performing simple what-if analyses by using transport planning software. The intention was to design relatively simple (easy to build) models to perform these what-if analyses.

Smart card data was mapped to the public transport network, resulting in passengers per line and number of boarding, alighting and transferring passengers per stop. Visualizing this data for each hour of day proved to be valuable when analysing the network (i.e. peak directions, distribution of demand over time and space). When analysing capacity in peak periods, distinguishing between half hour periods proved to be useful.

To construct OD demand matrices between stops, trips with one or more transfers in between are aggregated. A time threshold of 25–35 minutes appeared to be a good value to identify most transfers, while most short stay back-and-forth trips are still identified as separate trips. The OD demand matrix is assigned to the network to reproduce the measured passenger flows. Once the assignment can reproduce the passenger flows simple what-if analysis becomes possible. With fixed demand, line changes can be investigated. With the introduction of an elasticity method on the demand matrix, modal-split calculations are possible.

In our approach, we explicitly take comfort into account, since it is a relevant quality indicator, which is often neglected. We showed that the effect of a frequency increase in a congested public transport line in terms of additional passengers becomes significantly larger when comfort effects are included. We expect this effect to be closer to reality, because for choice travellers, crowding can be an important reason to choose an alternative mode. From a policy perspective this also indicates that benefits of such measures can be underestimated when comfort is not incorporated in the demand modelling framework. We also illustrated the potential of these models to be applied in practice, given the limited calculation times required.

The method described above was applied in a case study, being the tram network in The Hague. The tool turned out to be very valuable for the operator to gain insights into small changes. However, the approach has some limitations and shortcomings. First of all, the elasticity method is only valid for short term predictions and only unimodal (public transport) results are provided. We recommend further research on region specific elasticities. With the availability of smart cards, valuable revealed preference research is possible after changes in level of service. Another anticipated improvement is related to the zonal system. In this case the zones are at the stops making what-if analysis on stop choice rather limited. In an anticipated extension the smart card data station-to-station matrix will be converted to a proper zone matrix.

In addition to the presented application, several other application opportunities arise as well. We expect that smart card data will enable an increase in revealed preference research, thereby updating or adding new insights into elasticity values and modelling parameters. Useful insight may be gained from type of day, type of passenger and type of service area etc. We also see opportunities for data fusions applications. Combination of smart card and AVL data for instance, will provide more and better understanding of passenger reliability. The fusion of GSM and smart card data is promising – modal share per area and/or moment may be assessed quickly.

However we also face some challenges. There are concerns about availability of the data and privacy agreements that must be taken into account. Privacy is the most important issue, because individual data is used. Dutch privacy law states that processing individual data is not permitted and that data must not be preserved for more than 18 months. It also requires that before the start of research in which smart card data is used, the objective should be clearly stated. The dataset is not allowed to be used for other purposes.

Another concern is the availability for analysis. The data is owned by public transport operators and most of them see it as confidential company information. Combining data from several operators is a challenging topic. However an attempt of doing so is being made in The Netherlands in Nijenstein and Bussink (2015). This paper demonstrates how they created passenger journeys (consisting of multiple trips, thus including transferring) by combining smart card data of multiple operators.

## ACKNOWLEDGEMENTS

The authors are thankful for the data and tooling provided by HTM The Hague and Goudappel Coffeng.

## REFERENCES

Bagchi, M. and White, P. 2005. The Potential of Public Transport Smart Card Data. Transport Policy, Vol. 12, No. 5, pp. 464-474.

Balcombe, R., Mackett, R., Paulley, N., Preston, J., Shires, J., Titheridge, H., Wardman, M. and White, P. 2004. The demand for public transport: a practical guide.

Bouman, P., van der Hurk, E.L., Kroon, T., Li, and Vervest, P. 2013. Detecting activity patterns from smart card data. In 25th Benelux Conference on Artificial Intelligence (BNAIC 2013).

Brands, T., de Romph, E., Veitch T. and Cook, J. 2014. Modelling public transport route choice with multiple access and egress modes. Transportation Research Procedia, Vol. 1, pp. 12-23.

Bregman, S. 2012. Uses of Social Media in Public Transportation. Transit Cooperative Research Program (TCRP) Synthesis 99. Transportation Research Board, Washington.

Cats O., Larijani, A.N., Ólafsdóttir, A., Burghout, W., Andreasson I. and Koutsopoulos, H.N. 2012. Holding Control Strategies: A Simulation-Based Evaluation and Guidelines for Implementation. Transportation Research Record 2274, pp. 100-108.

Cepeda, M., Cominetti, R. and Florian, M. 2006. A frequency-based assignment model for congested transit networks with strict capacity constraints: characterization and computation of equilibria. Transportation Research Part B: Methodological, 40, pp. 437–459.

Cheung, F. 2006. Implementation of Nationwide Public Transport Smart Card in the Netherlands: Cost-Benefit Analysis. Transportation Research Record, Journal of the Transportation Research Board, No. 1971, Transportation Research Board of the National Academies, Washington, D.C., pp. 127-132.

Douglas Economics. 2006. Value and Demand Effect of Rail Service Attributes. Report to RailCorp. Wellington, New Zealand.

Florian, M. 2002. Frequency based transit route choice models. Chapter 6 in: Advanced Modeling for Transit Operations and Service Planning, ed. William H.K. Lam, Michael G. H. Bell.

Furth, P.G., Hemily, B., Muller, T.H.J. and Strathman, J.G. 2006. TCRP Report 113: Using Archived AVL-APC Data to Improve Transit Performance and Management. Washington, D.C.

Hasan, S., Schneider, C.M., Ukkusuri, S.V. and Gonzalez, M.C. 2013. Spatiotemporal patterns of urban human mobility, J. of Statistical Physics, Vol. 151 (1-2), pp. 304-318.

Hickman, M. 2004. Evaluating the Benefits of Bus Automatic Vehicle Location (AVL) Systems, in: D. Levinson and D. Gillen (eds.), Assessing the Benefits and Costs of Intelligent Transportation Systems, Chapter 5, Kluwer, Boston.

Joksimovic, D. and van Grol, R. 2012. New Generation Dutch National And Regional Models — An Overview Of Theory And Practice, European Transport Conference.

Kurauchi, F., Schmöcker, J.D., Shimamoto, H. and Hassan, S.M. 2014. Variability of commuters' bus line choice: an analysis of oyster card data. Public Transport, Vol. 6, No. 1-2, pp. 21-34.

Lee, A., van Oort, N. and van Nes, R. 2014. Service reliability in a network context, Transportation Research Record, No. 2417, pp. 18-26.

Ma, X., Wu, Y.J., Wang, Y., Chen, F. and Liu, J. 2013. Mining smart card data for transit riders' travel patterns. Transportation Research Part C: Emerging Technologies, Vol. 36, pp. 1-12.

Morency, C., Trepanier, M. and Agard, B. 2007. Measuring transit use variability with smart-card data. Transport Policy, Vol. 14, No. 3, pp. 193-203.

Munizaga, M. and Palma, C. 2012. Estimation of a disaggregate multimodal public transport Origin-Destination matrix from passive smart card data from Santiago, Chile. Transportation Research C, Vol. 24, pp. 9-18.

MVA Consultancy. 2008. Valuation of Overcrowding on Rail Services. Prepared for Department for Transport.

Nijenstein, S. and Bussink, B. 2015. Combining multimodal smart card data, Presented at European Transport Conference, Frankfurt.

Neema, N., Hickman, M. and Ma, Z-L. 2015. Activity detection and transfer identification for public transport fare card data, Transportation.

Oostra, R. 2004. Elasticiteitsonderzoek binnen het vervoergebied van HTM. TU Delft (In Dutch).

Park, J., Kim, D.J. and Lim, Y. 2008. Use of Smart Card Data to Define Public Transit Use in Seoul, South Korea. Transportation Research Record, Journal of the Transportation Research Board, No. 2063, Transportation Research Board of the National Academies, Washington, D.C., pp. 3-9.

Pel, A.J., Bel, N.H. and Pieters, M. 2014. Including passengers' response to crowding in the Dutch national train passenger assignment model. Transportation Research Part A: Policy and Practice, 66, pp. 111-126.

Pelletier, M., Trepanier, M. and Morency, C. 2011. Smart card data use in public transit: A literature review. Transportation Research Part C: Emerging Technologies, Vol. 19, No. 4, pp. 557-568.

Schmöcker, J.D., Fonzone, A., Shimamoto, H., Kurauchi, F. and Bell, M.G.H. 2011. Frequency-based transit assignment considering seat capacities. Transportation Research Part B: Methodological, 45(2), pp. 392-408.

Schmöcker, J.D., Shimamoto, H. and Kurauchi, F. 2013. Generation and calibration of transit hyperpaths. Transportation Research Part C: Emerging Technologies, Vol. 36, pp. 406-418.

Seaborn, C., Attanucci, J. and Wilson, N.H.M. 2009. Analysing Multimodal Public Transport Journeys in London with Smart Card Fare Payment Data. Transportation Research Record, Journal of the Transportation Research Board, No. 2121, Transportation Research Board of the National Academies, Washington, D.C., pp. 55-62.

Significance, VU University, John Bates Services, TNO, NEA, TNS NIPO and PanelClix. 2013. Values of time and reliability in passenger and freight transport in The Netherlands. Report for the Ministry of Infrastructure and the Environment, Significance, The Hague.

Litman, T. 2013. Transport Elasticities: Impacts on Travel Behaviour. Sustainable Urban Transport Technical Document, GIZ.

Transportation Research Board. TCRP REPORT 95. 2004. Transit Scheduling and Frequency Traveler Response to Transportation System Changes. Chapter 9.

Trépanier, M., Tranchant, N. and Chapleau, R. 2007. Individual trip destination estimation in a transit smart card automated fare collection system. Journal of Intelligent Transportation Systems, Vol. 11, pp. 1-14.

Van den Heuvel, J., Voskamp, A., Daamen W. and Hoogendoorn, S.P. 2015. Using bluetooth to estimate the impact of congestion on pedestrian route choice at train stations, Traffic and Granular flow '13, M. Charaibi et al. (eds), Switzerland.

Van Oort, N., Boterman, J.W. and van Nes, R. 2012. The impact of scheduling on service reliability: trip-time determination and holding points in long-headway services. Public Transport, 4(1), pp. 39-56.

Van Oort, N. and van Nes, R. 2009. Line length versus operational reliability: network design dilemma in urban public transportation. Transportation Research Record, No. 2112, Washington, D.C., pp. 104-110.

Van Oort, N., Brands, T., de Romph, E. 2015a, Short-Term Prediction of Ridership on Public Transport with Smart Card Data, *Transportation Research Record*, No. 2535, pp. 105-111.

Van Oort, N., Brands, T., de Romph, E. and Flores, J.A. 2015b. Unreliability effects in public transport modelling, International Journal of Transportation Vol. 3, No. 1, pp. 113-130.

Van Oort, N., Sparing, D., Brands, T. and Goverde, R.M.P. 2015c. Data driven improvements in public transport: the Dutch example, *Public Transport*, Vol. 7(3), pp. 369-389.

Van Oort, N., Yap, M.D. and Oud, M. 2016. Understanding public transport passenger behaviour during (un)planned disturbances: insights from smartcard data. To be presented at the Word Conference on Transport Research Society, China.

Wang, W., Attanucci, J.P. and Wilson, N.H.M. 2011. Bus Passenger Origin-Destination Estimation and Related Analyses. Journal of Public Transportation, Vol. 14, No. 4, pp. 131-150.

Wardman, M. 2004. Public transport values of time. Transport Policy, Vol. 11, No. 4, pp. 363-377.

Wardman, M. 2012. Review and meta-analysis of U.K. time elasticities of travel demand. Transportation, Vol. 39, No. 3, pp. 465-490.

Wardman, M. and Whelan, G. 2011. Twenty Years of Rail Crowding Valuation Studies: Evidence and Lessons from British Experience. Transport Reviews 31(3), pp. 379-398.

## AUTHOR BIOGRAPHY

**Niels van Oort** works as an assistant professor Public Transport at Delft University of Technology and via his job as a public transport consultant at Goudappel Coffeng he is involved in several public transport projects. His main fields of expertise are public transport planning, dealing with the passenger perspective, service reliability and Big Data. He finished his PhD on service reliability in 2011. Niels develops approaches and tools to transfer data into knowledge with the aim to improve public transport. His research results are available at: https://nielsvanoort.weblog.tudelft.nl/

**Ties Brands** obtained his two MSc degrees from University of Twente in 2008 and works at Goudappel Coffeng as a public transport consultant, specialized in public transport modelling and (smartcard) data analysis. In 2015 he finished a PhD project on public transport network optimization. He works on projects such as network planning studies, ridership predictions, cost benefit analyses and data analyses. In recent years, this has included analyses of smart card data from several Dutch cities and regions.

**Erik de Romph** studied Computer Science and received a PhD in Transport Planning at Delft University of Technology in The Netherlands. Erik is one of the founders of the software package for transport planning: OmniTRANS. He was managing director of Omnitrans International from 2007 to 2013. In 2013 he accepted a chair as professor in transport modelling in Delft. Erik has been involved in various research projects using new (big-) data sources in transport modelling, such as cell-phone data and smart card data.

**Menno Yap** is working as consultant public transport at Goudappel Coffeng. Menno graduated cum laude from the Delft University of Technology on the topic 'robustness of public transport networks from a passenger perspective'. At Goudappel Coffeng, Menno is involved in applying and improving (data driven) public transport models. He also focuses on the role of innovative transportation systems, like automated vehicles, in public transport. He recently started a PhD research on optimisation of transfers.

Chapter 12

# Assessment of Traffic Bottlenecks at Bus Stops

**K. Makimura[1,*], T. Nakamura[2], T. Ishigami[1] and R. Imai[3]**

**ABSTRACT**

This chapter describes the methodology used in Saitama, Japan to tackle traffic bottlenecks in the vicinity of bus stops. Two different kinds of tracking data—probe car data and bus smart card data—were utilized to study which infrastructure improvements around bus stops are most in need. Probe car data are used to assess where buses block the traffic, while smart card data help to understand the temporal distribution of passenger demand. It has been found that a simple analysis of smart card data can be a useful and powerful tool in assisting local authorities to strategically decide where to invest their resources, limited as they are.

## 1. INTRODUCTION

In recent years, digitized tracking data captures an enormous amount of human movement, in real time. The data includes smart card data from buses and railway systems, as well as probe person data and probe car data, which are collected by GPS-enabled mobile phones providing position identification and car navigation systems. Such tracking data cover wide areas with great efficiency 24 hours per day and 365 days per year (Kawasaki and Hato, 2004).

Smart card data is more easily available to local authorities and various transport planning authorities. The contributors in this book discuss several advanced methods for analyse of smart card data at the disaggregate level. However, detailed analysis of individual travel patterns

[1] Institute of Behavioral Sciences, 2-9 Ichigayahonmura-cho, Shinjuku-ku, Tokyo, Japan. Email: kmakimura@ibs.or.jp

[2] Department of Urban Management, Graduate School of Engineering, Kyoto University, Japan.

[3] Department of Urban and Civil Engineering, Tokyo City University, Japan.

* Corresponding author

are often not done for practical applications as aggregate-level data are preferred. There are many useful applications of the aggregate-level smart card data (Bagchi and White, 2005 and Pelletier et al., 2011). From the standpoint of practical application, this study focuses on smart card data only used at the aggregate level.

There is by now a significant amount of literature on understanding travel patterns through digitized tracking (e.g. Akiyama et al., 2011). In previous research though mostly a single tracking data source was used, but in the light of current advances in the collection of data from various trails, it is possible now to create a combined-analysis which uses features of data from multiple different trails and to disseminate new knowledge.

Turning to examples of the application of digitized tracking data in road administration, tracking data from ordinary vehicles (hereinafter called, 'probe car data') are used to grasp traffic behavior and conditions for drafting transportation planning and measuring the effects of road maintenance and improvement (Momma et al., 2011). Currently, only probe car data have been used, but road administrations need to understand the bus stop traffic blockage points (Kinuta et al., 2008 and Makimura et al., 2010). That is, at which bus stops the traffic is often blocked due to buses requiring time to load the waiting passengers needs to be studied. Furthermore, the authors exchanged opinions with road administrators of local governments, when they faced issues, such as difficulty in acquiring a comprehensive picture of the situation because carrying out field surveys at every bus stop requires a large volume of work. Hence they confirm an overpowering need for efficient extraction of such 'bus stop traffic blockage points'. For this smart card data turn out to be of practical use as will be explained in this chapter.

The objective of this contribution is, therefore, to jointly look at probe car data and smart card data in order to establish support measures for improving traffic movement in the vicinity of bus stops. The methodology is applied to evaluate bus stops in Saitama City, Japan, where both smart card and vehicle probe data are available.

In Section 2 we summarize previous research and discuss the positioning of this study. In Section 3 we discuss how the evaluation indices are constructed that support decisions to invest in measures that improve traffic in the vicinity of bus stops. In Section 4 we verify the usefulness of the proposed support method by a case study in Saitama City. In Section 5 we summarize and provide an outlook to future work.

## 2. BACKGROUND OF THIS STUDY

Previous research on probe car data includes consideration of the accuracy of the data acquired and the required sample sizes (Ishida et al., 2001), routing based on technology for matching collected data to maps (Li et al., in press), evaluation of traffic condition from the obtained data (Tamiya and Seo, 2002) and measurement of effects on road maintenance and improvement (Momma et al., 2011). Consideration of the accuracy of massive probe car data and issues connected to required sampling size are summed up in particular in the research of Hashimoto et al. (2014).

Previous research on smart card data includes elucidation of traffic behavior and use for traffic surveys utilizing the characteristics of long-term data, use in demand estimation and in transportation planning as discussed in various chapters in this book. Also, in Japan, research on smart card data has been conducted for more than 10 years. Okamura et al. (2002) examine the potential of ridership data from the common magnetic cards used in the Hiroshima metropolitan area to complement or replace the existing survey data. Nogami et al. (2011) use smart card data from Kochi prefecture and elucidate the public transportation usage by OD analysis of the number of trips made per zone and analysis of transfers between bus and streetcar. Nakajima et al. (2009) analyse changes in user in-vehicle time due to the opening of a new railway in Osaka to gauge the ramifications of its opening. Yabe and Nakamura (2008) use smart card data from the Tokyo area to analyse the relationships between card dissemination rate and bus dwell time, and between reducing dwell time and operating hours. Finally, Makimura et al. (2010) initiate the discussion on applying smart card data for planning improvements in the bus stop facility, which is also the objective of this study.

In summary, previous research using probe car data focused on verifying the accuracy of collected data and on understanding the characteristics of the data. Examples of its use to measure the effects of road maintenance and improvement are also found in literature. Similarly, with environments enabling the collection of large amounts of data over long periods bus smart card data could be prepared, and verification of the effects of the introduction of smart cards on routes and the opening of new routes became possible in previous research.

Those examples used one type of tracking data for analysis. Combining multiple types of tracking data from a single area potentially enables further detailed analysis to support public transportation planning.

## 3. DEVELOPMENT OF EVALUATION MEASURES

The bus stop improvement program based on the Niigata City Omnibus Town Plan is an example of a conventional study of bus facility improvement (Niigata City, 2007). Under the program a field survey of each bus stop was conducted and bus stops (especially candidates for the addition of roofs) were selected for improvement of their service level in terms of routes and numbers of buses, barrier-free priority zones, roofing as well as considerations of access to public facilities such as hospitals.

Compared to the Niigata study, the important feature of this chapter is that evaluation measures are obtained through tracking data which were previously difficult to obtain, such as bus travel speed, number of users and travel speed of ordinary vehicles.

## 3.1 Procedure for Obtaining Evaluation Measures

Figure 1 outlines the procedure to obtain evaluation measures using probe car and smart card data. The details of each step are explained in the following pages.

Step 1 Sorting bus stops by usage

↓

Step 2 Analysis of 'dynamic conditions' at bus stops
Confirmation of usage status and travel performance using smart card and private-sector probe data for bus stops educed in Step 1 above

↓

Step 3 Analysis of 'static conditions' at bus stops
- Service level of buses stopping at the stop (number of buses, lines, etc.)
- Availability of a roof and seats at the stop?
- Have Public Transport priority measures been adopted?

↓

Step 4 Extraction of traffic blockage points near the bus stops

↓

Step 5 Creation of a chart for each bus stop

**Fig. 1.** Procedure for study of needs in bus stop facilities and surrounding road infrastructure

*Step 1: Compilation of the number of users at each bus stop*

Use of smart card data from buses to collect the number of users at each bus stop. Through smart card data large numbers of bus stops can be analysed, even if there are more than 1,000 bus stops as is the case in larger cities.

*Step 2: Obtaining 'dynamic conditions' at bus stops*

Utilization of probe car data and smart card data from buses to set indicators according to the predefined objectives and to obtain information about variations in traffic usage as per the time of the day, the day of the week or the weather conditions. Accordingly, the nine indicators shown in Table 1 are set.

The average travel speed during the interval is calculated from bus smart card data as per the following procedure:

As data from smart cards are collected at the time of getting on and off trains/buses when passengers touch the equipment, then those timings are recorded as travel time between bus stops. On the other hand, link travel time data are obtained by converting the calculated travel time between bus stops to travel time by link. The method of conversion is indicated below.

First, when users touch the equipment with their smart cards, the travel time between bus stops is collected and so are the individual ID and time recorded. Thus, data are recorded at payment in areas with uniform bus fare, and at the time of getting on and off in the areas with distance-based bus fares. The required travel time between bus stops is calculated with the assumption that the time recorded on the equipment is the time the bus arrived at the stop where the passengers were waiting (Figure 2).

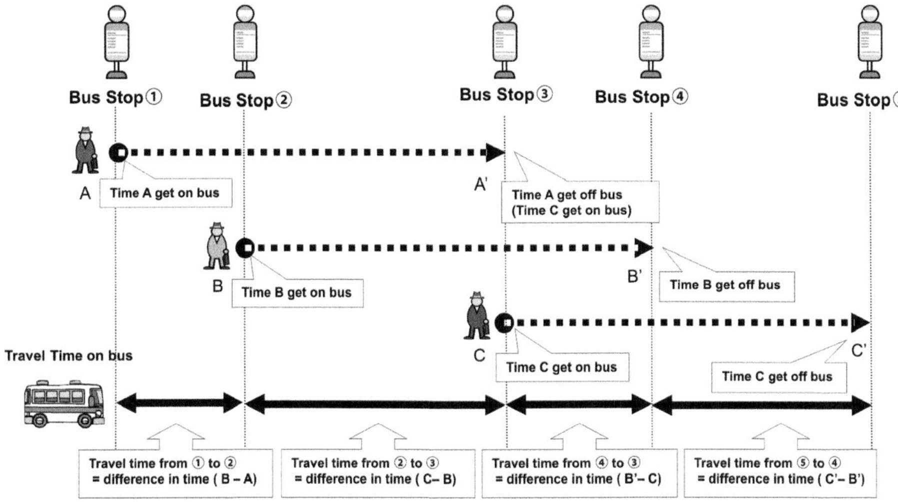


**Fig. 2.** Estimation method of the travel time between bus stops

As data of multiple cards are recorded at bus stops, the first record is considered the time of arrival, and the last record as time of departure. This helps in estimating the travel time while excluding the stopping time at bus stops. If there is any bus stop with no passenger, data at the nearest bus stops with records are used and proportionally divided by the distance between the bus stops to estimate the travel time.

Next, the method of conversion of travel time between bus stops into travel speed by road link is explained. Simply put, by proportionally dividing the travel speed between the bus stops into road links according to distance, the travel speed is convertedinto link travel times at 10-minute and 60-minute intervals. When doing so, if data of multiple bus routes are available, then the bus travel time is considered the average travel time between the bus stops for 10 minutes and 60 minutes, respectively. This is explained in Figure 3. Considering the location of the bus stops on a link below, travel time of the link between Node Y and Node Z in Figure 3 uses the travel time between bus stop A and bus stop B ($T_{AB}$) and travel time between bus stop B and bus stop C ($T_{BC}$). Because of the location of Node Y, $T_{AB}$ is divided into distance $D_{AY}$ and distance $D_{YB}$. Then, they are proportionally divided according to distance and estimated using the equation $T_{YB} = T_{AB} * (D_{YB}/D_{AY} + D_{YB})$. Similarly, $T_{BZ}$ is estimated as $T_{BC}$ $*(D_{BZ}/D_{BZ}+D_{ZC})$. In doing so, travel time between node Y and node Z would be $T_{YB} + T_{BZ}$, and the travel speed is obtained by dividing the result with $D_{YZ}$.

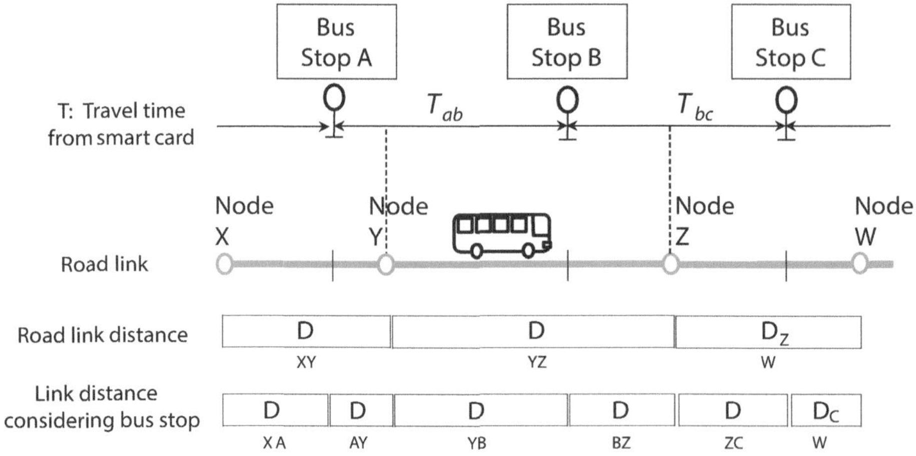


**Fig. 3.** Conversion of travel time between bus stops to travel time between links

The data in Table 1 helps to distinguish bus stop usage rates during different time periods of the day, different weekdays as well as for bus stops located in different types of land-use areas. Moreover, usage depending on weather conditions can be analysed, such as the impact of rainy days versus clear days. Further, the analysis of probe vehicle data helps to determine the bus/ordinary vehicle travel performance around bus stops. This is important since the traffic performance in one location can vary widely, depending on the time of day. We emphasize once again that the indicators set for Step 2 are required but are difficult to capture by conventional bus stop improvement programs. Using probe data and smart card as described in this chapter simplifies this task significantly.

*Step 3: Static travel conditions around bus stops*

The bus service level in terms of the number of buses and routes servicing at the bus stop, the number of lanes on the road where the bus stop is located and the state of bus bay facility are evaluated. Further, the state of the Public Transportation Priority System, the introduction of the priority (dedicated) lanes as well as the existence of bus-stop roofs and seats are set as indicators. Table 2 summarizes the nine indicators for 'static conditions'.

While the dynamic conditions of Step 2 can be obtained through tracking data, field surveys are required for obtaining the static conditions. However, because bus stops with high ridership are already selected at Step 1, rather than a field survey of every bus stop as was done in the case of Niigata City, field surveys can instead be targeted at potentially critical bus stops with higher ridership.

*Step 4: Analysis of traffic blockage points near bus stops*

Traffic blockage points in the vicinity of bus stops refer to spots where buses stop to let passengers board or alight and in the process, block the

**Table 1.** Indicators of Dynamic Conditions of Bus Stops

| | Category | Indicator | Details |
|---|---|---|---|
| Organization of dynamic conditions | Bus usage situation | Average number of users | Daily average number of users at the stop for weekdays, Saturdays, and Sundays/holidays as calculated from smart cards |
| | | Variation in number of users | Number of users at the stop according to smart cards, organized by month |
| | | Usage rate during commuting period (weekdays) | Usage percentage during weekday commuting period (7:00 and 8:00 hours) is calculated as "number of commuter users (7:00 and 8:00 hours) / weekday number of users" |
| | | Sunday/holiday usage rate | Usage percentage on weekends and holidays compared with weekdays is calculated as "number of Saturday users (persons/day) / number of weekday users (persons/day)" |
| | | Rainy weather usage rate | Calculated for weekdays, Saturdays, and Sundays/holidays as "number of rainy-day users (persons/day) / number of clear-day users (persons/day)" |
| | Bus/ordinary vehicle travel performance | Average bus travel speed | Average bus travel speed for the stop is calculated for the weekday morning peak (7:00 and 8:00 hours), middle of the day (9:00 to 16:00 hours) and evening peak (17:00 and 18:00 hours) |
| | | Variation in average bus travel speed | Calculate variations in average weekday travel speed by month and in average travel time by time of day |
| | | Average travel speed of ordinary vehicles | Calculate average travel speed of ordinary vehicles immediately before the stop for morning peak, middle of the day, evening peak and daily average |
| | | Variation in average travel speed of ordinary vehicles | Calculate variations in average weekday travel speed by month and in average travel time by day time |

**Table 2.** Indicators of Static Conditions at Bus Stops

| | Category | Indicator | Details |
|---|---|---|---|
| Organization of static conditions | State of bus operation | Service provider | Bus service provider using the stop |
| | | Servicing routes | Number of routes that service the stop |
| | | Servicing buses | Organized by numbers of buses servicing the stop on weekdays, Saturdays, and Sundays/holidays |
| | Bus operation environment | Number of lanes | Number of lanes at the stop's location |
| | | State of bus bay development | Is a bus bay provided at the stop? |
| | | State of roof development | Is a roof provided at the stop? |
| | | State of seating development | Is seating provided at the stop? |
| | | State of PTPS introduction | Are routes servicing the stop PTPS routes? |
| | | State of priority (dedicated) lane adoption | Do routes servicing the stop have priority (dedicated) bus lanes? |

subsequent traffic, creating congestion, as illustrated in Figure 4. Extraction of traffic blockage points is based on results from probe car data and smart card data plotted on digital road maps (hereinafter called, 'DRMs'). The probe car data are used to generate and compare travel speeds of ordinary vehicles immediately before and after the bus stops.

More specifically, the rate of traffic blockage occurrence is defined by using the three travel speeds defined below and as shown in Figure 5.

**Fig. 4.** Traffic blockage occurring at a bus stop

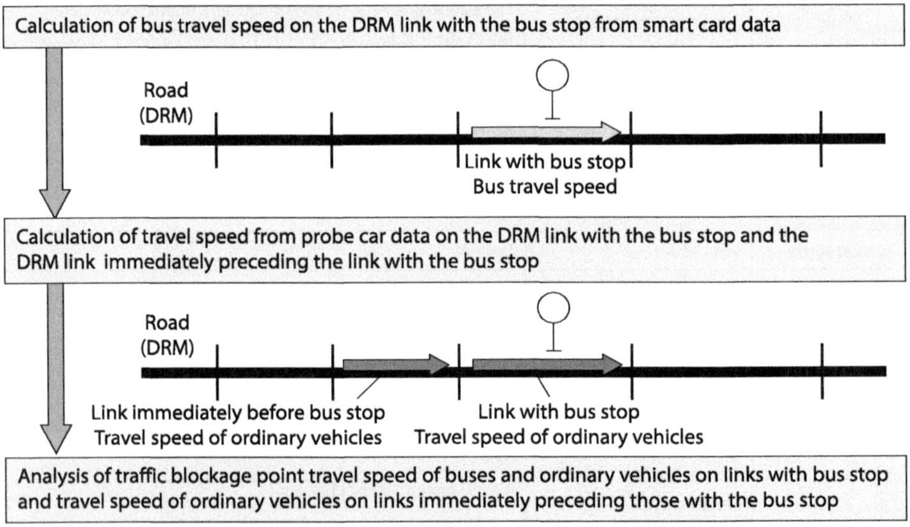


**Fig. 5.** Travel speed used to calculate rate of traffic blockage occurrence

$V_{t,i}^{bus}$ : Travel speed of bus at time $t$ on link $i$ at which the bus stop of interest is located.

$V_{t,i}^{car}$ : Travel speed of ordinary vehicles at $t$ time on link $i$ at which the bus stop of interest is located.

$V_{t,i-1}^{car}$ : Travel speed of ordinary vehicles at time $t$ on link $(i–1)$ immediately before the bus stop.

Using the three defined travel speeds, two conditions are set (see Figure 6).

$$V_{t,i}^{car} - V_{t,i}^{bus} < 0 \tag{1}$$

$$V_{t,i}^{car} - V_{t,i-1}^{car} < 0 \tag{2}$$

The above conditions show ordinary vehicles having a travel speed lower than that of buses on the bus stop link (Eq. 1) and an even lower travel speed on the link preceding the bus stop link (Eq. 2). Let $n$ be the number of travel speed data samples taken at 15-minute intervals for which Eq.1 holds. The rate of traffic blockage occurrence at the bus stop is then defined as follows.

$$Blockage\ Rate = n/N \tag{3}$$

This is illustrated in Figure 7. A large number of plots in the shaded lower left area of the figure indicates that ordinary vehicles are unable to pass buses at the stop, meaning that traffic blockages occur.

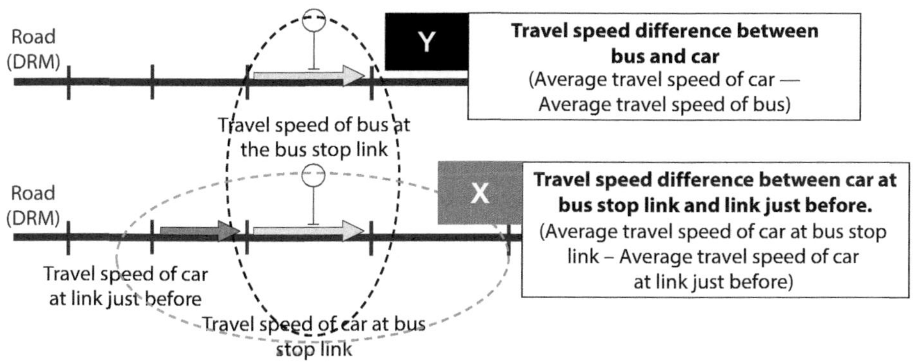


**Fig. 6.** Conditional expressions using the three travel speeds

*Step 5: Creation of a chart for each bus stop*

In the final analysis, effort is made to display the information gathered in steps 1 to 4. For easy interpretation and discussion with decision makers it was found relevant to use what in Japan are known as 'hospital-type charts' as shown in Figure 13 below in our case study that is discussed next.

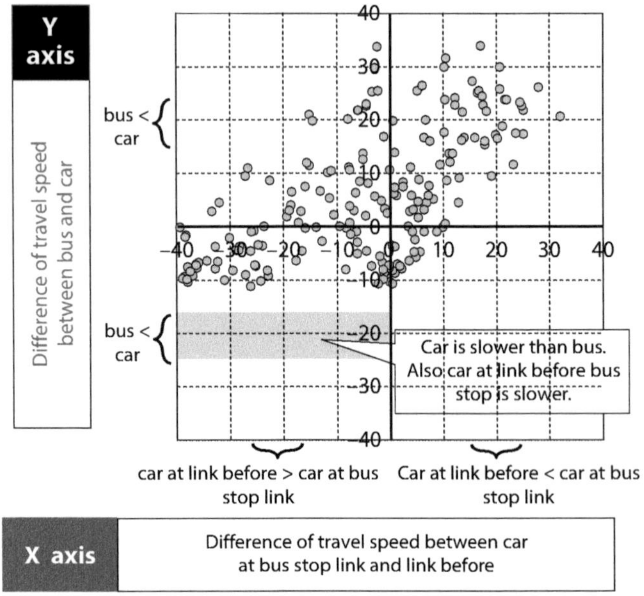


**Fig. 7.** Image of organizing the rate of traffic blockage occurrence

## 4. SAITAMA CITY CASE STUDY

This section demonstrates the usefulness of the above-discussed methodology. The method is applied to all bus stops in Saitama City which are served by buses that take smart card payments. The study was carried out at the advice of the Omiya National Highway Office of the Kanto Regional Development Bureau and public transportation related divisions of Saitama Prefecture and Saitama City.

### 4.1 Saitama City

Saitama City has a Transport Strategy Council which has been carrying out a study aimed at developing an urban transport strategy since 2009. The strategy aims to achieve improved travel speeds and punctuality with better mobility between local hubs and neighboring cities. Buses are expected to play a major role in this.

As shown in Figure 8, Saitama City's mobility depends crucially on north-south railways that connect the city to Tokyo. However, the railway network running east to west from the major train stations Omiya Station and Urawa Station are underdeveloped, leading to a dependence on buses. There is, therefore, a great desire to improve the bus operating environment and the bus stops. With over 1,000 bus stops in the city, no discussion was begun with the Transport Strategy Council, where to begin and how to improve the travel speeds. Given this background, Saitama City has been chosen as the target area for this case study.

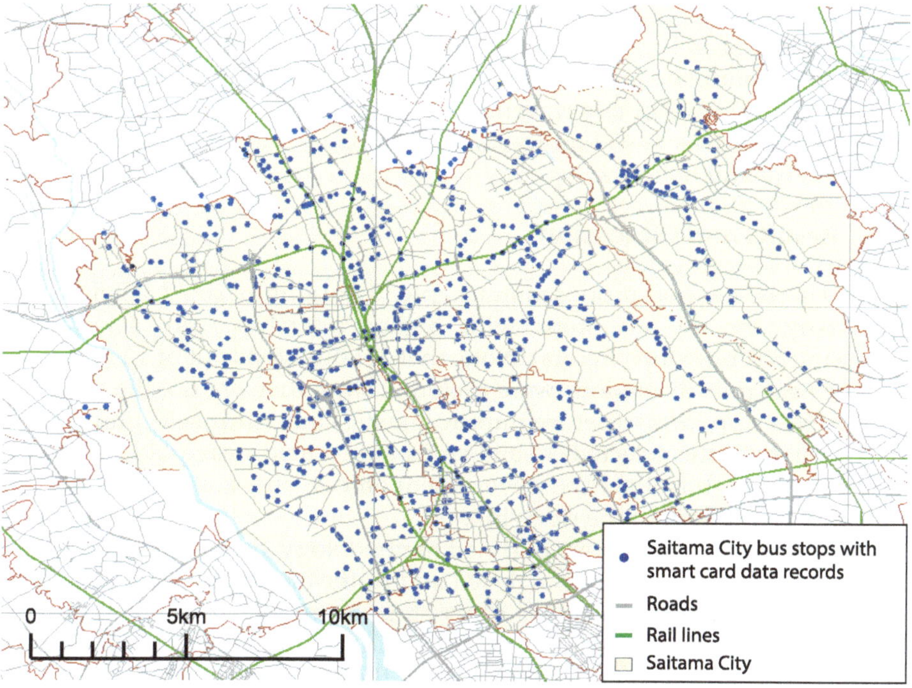


**Fig. 8.** Analysis area (Saitama City) and bus stops

## 4.2 Overview of Tracking Data Use

The study area and analysis period are shown in Table 3. Considering the bus operation hours, the data analysis period is from 6 a.m. to 10 p.m. and both probe car and smart card data have been collected:

**Table 3.** Study Area and Analysis Period

| Category | Details |
|---|---|
| Area | Saitama City, Saitama Prefecture |
| Analysis period | June 2010 (one month)<br>Weekdays: 22, Saturdays: 4, Sundays/holidays: 4 |

a) *Probe car data*

The probe car data used for this study were collected from car navigation systems installed in ordinary vehicles. Private-sector businesses provided travel time data collected at five-minute intervals from DRM links. Travel speed for each DRM link was calculated by using the travel time data and DRM link extension.

b) *Smart card data from buses*

The smart card data used for this study were from 'Suica' and 'PASMO', the two major smart cards that are used within the larger

Tokyo metropolitan region and have been in operation since March 2007. As of February 2010, about 65 million records were collected every month. In June 2010, the month selected for the analysis, the average number of bus travellers using one of these smart cards in the target area was 41,659 on weekdays, 25,335 on Saturdays and 18,619 on Sundays and holidays.

## 6.3 Results and Discussion

Results of the method proposed in Chapter 3 are summarized as follows.

Firstly, the 30 bus stops with higher ridership were selected by analysing 802 bus stops in total. Note, that 314 bus stops out of 1,116 that are located near railway stations were excluded since they are not the target for road or bus stop infrastructure improvements. Under this selection criteria, unsurprisingly the selected 30 bus stops are located near public facilities, such as schools, hospitals and government offices. These 30 stops accounted for about 27 per cent of the ridership.

Following the previously outlined methodology, the dynamic and static conditions of the stops were analysed and traffic blockage points identified. Only the previously selected 30 bus stops were evaluated. The authors are of the view that it is advisable to decide the number of bus stops to be analysed based on conditions, such as the size of the analysis area, the percentage of bus stops and the overall percentage ridership.

As an example of the dynamic conditions, Table 4 presents some indicators of the 30 bus stops and Figure 9 presents an in-depth analysis of the Daitakubo bus stop. The figure describes in particular the decrease of usage during weekends, thus providing valuable information for understanding the importance of the bus stop as well as the potential impact of bus diversions during weekends for infrastructure improvements.

For the analysis of blockage points the traffic speed indicators shown in Table 5 and Figure 10 are important. At the Daitakubo bus stop the travel speed is low during weekday mornings and peak in the evenings. The travel speed of ordinary vehicles is about the same as that of buses during evening peak hours on weekdays.

**Table 4.** Dynamic Indicators

| Rank | Name of Bus Stop | Av. number of passengers (person/day) | | | Rate of commuter time | Rate of Sunday | | Rate of Rainy day | | |
|---|---|---|---|---|---|---|---|---|---|---|
| | | weekday | Saturday | Sunday | weekday | Saturday | Sunday | weekday | Saturday | Sunday |
| 1 | Saimata Unv | 1,198 | 594 | 601 | 0.12 | 0.50 | 1.01 | 1.32 | 1.13 | 0.69 |
| 2 | Daitakubo | 638 | 359 | 271 | 0.38 | 0.56 | 0.75 | 1.36 | 1.08 | 1.09 |
| 3 | Jichiidaiiryo | 561 | 133 | 75 | 0.05 | 0.24 | 0.56 | 1.27 | 1.58 | 1.02 |
| 4 | Segasaki | 494 | 287 | 202 | 0.53 | 0.58 | 0.70 | 1.40 | 1.15 | 1.11 |
| 5 | Shiritsubyouin | 492 | 201 | 161 | 0.15 | 0.41 | 0.80 | 1.24 | 0.85 | 1.04 |
| 6 | Kyoikusentamae | 441 | 258 | 163 | 0.35 | 0.58 | 0.63 | 1.26 | 0.93 | 1.12 |
| 7 | Horinouchibashi | 396 | 337 | 282 | 0.38 | 0.85 | 0.84 | 1.31 | 0.93 | 0.75 |
| 8 | Higashisegasaki | 394 | 281 | 181 | 0.49 | 0.71 | 0.65 | 1.41 | 1.19 | 1.20 |
| • | • | • | • | • | • | • | • | • | • | • |
| • | • | • | • | • | • | • | • | • | • | • |
| • | • | • | • | • | • | • | • | • | • | • |
| 27 | NegishiGotyoume | 275 | 171 | 111 | 0.41 | 0.62 | 0.65 | 1.40 | 1.10 | 0.89 |
| 28 | Kenchomae | 306 | 63 | 42 | 0.17 | 0.20 | 0.67 | 1.48 | 1.31 | 1.55 |
| 29 | Hosono | 272 | 173 | 113 | 0.50 | 0.63 | 0.65 | 1.45 | 1.08 | 1.08 |
| 30 | Eiwakitamachi | 266 | 176 | 124 | 0.43 | 0.66 | 0.70 | 1.28 | 0.97 | 0.92 |

**Table 5.** Travel Speed at Bus Stops

| Rank | Name of Bus Stop | Average travel speed of Buses | | | | | | Average travel speed of Cars | | | | | |
|---|---|---|---|---|---|---|---|---|---|---|---|---|---|
| | | weekday | | | | Saturday | Sunday | weekday | | | | Saturday | Sunday |
| | | Am Peak | Daytime | PM Peak | Day Ave | Day Ave | Day Ave | Am Peak | Daytime | PM Peak | Day Ave | Day Ave | Day Ave |
| 1 | Saimata Unv | 15.3 | 15.0 | 14.4 | 14.8 | 14.0 | — | 17.2 | 18.2 | 19.9 | 21.0 | 21.3 | 21.4 |
| 2 | Daitakubo | 13.0 | 14.4 | 13.0 | 14.3 | 13.9 | 14.7 | 20.0 | 18.1 | 12.2 | 21.9 | 18.8 | 17.3 |
| 3 | Jichiidaiiryo | 12.6 | 13.0 | 13.3 | 13.1 | 13.1 | 13.9 | — | — | — | — | — | — |
| 4 | Segasaki | 14.1 | 15.6 | 15.7 | 16.0 | 15.7 | 15.6 | 28.8 | 24.5 | 22.8 | 26.3 | 26.4 | 24.8 |
| 5 | Shiritsubyouin | 13.2 | 14.1 | 13.5 | 14.1 | 13.9 | 14.8 | 27.8 | 31.1 | 30.2 | 33.1 | 33.0 | 38.5 |
| 6 | Kyoikusentamae | 12.9 | 14.0 | 12.4 | 14.1 | 14.3 | 15.5 | 30.0 | 24.3 | 27.9 | 29.7 | 28.1 | 32.4 |
| 7 | Horinouchibashi | 14.3 | 16.1 | 15.1 | 16.3 | 17.1 | 17.7 | 28.1 | 28.2 | | 30.9 | 31.2 | 39.6 |
| ⋮ | ⋮ | ⋮ | ⋮ | ⋮ | ⋮ | ⋮ | ⋮ | ⋮ | ⋮ | ⋮ | ⋮ | ⋮ | ⋮ |
| 26 | Hrayamasantyome | 15.1 | 14.9 | 13.5 | 15.3 | 14.3 | 15.1 | 24.0 | 18.4 | 16.1 | 25.8 | 23.7 | 24.3 |
| 27 | NegishiGotyoume | 14.5 | 15.0 | 14.3 | 15.4 | 15.3 | 15.9 | 29.7 | 27.1 | 25.9 | 32.8 | 30.2 | 31.8 |
| 28 | Kenchomae | 13.2 | 13.8 | 14.2 | 14.4 | 14.7 | 15.5 | 19.6 | 19.9 | 19.6 | 20.5 | 21.9 | 22.2 |
| 29 | Hosono | 15.2 | 16.4 | 15.9 | 16.3 | 16.2 | 16.4 | 36.5 | 31.1 | 29.8 | 35.9 | 35.1 | 33.5 |
| 30 | Eiwakitamachi | 13.3 | 13.8 | 13.0 | 13.7 | 13.1 | 13.9 | 21.8 | 21.0 | 19.5 | 24.5 | 21.8 | 21.2 |

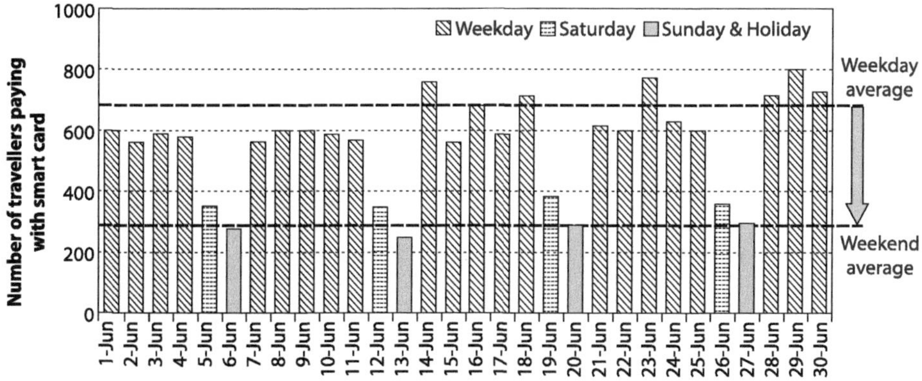


**Fig. 9.** No. of Users at Daitakubo Bus Stop by Weekday, Saturday and Sunday/Holiday

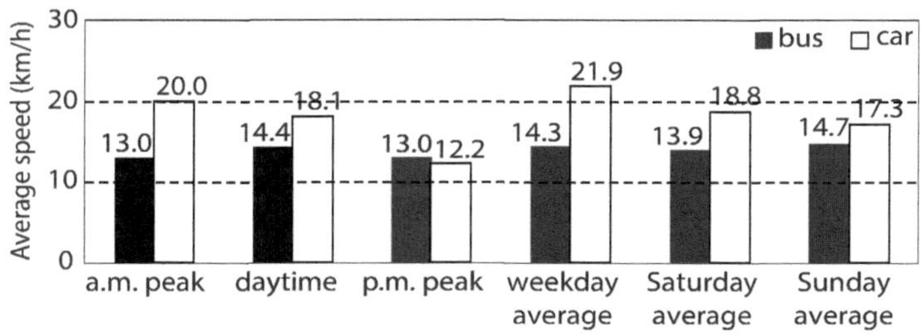


**Fig. 10.** Average Travel Speed by Time Segment at Daitakubo Bus Stop

Using these speed data, the rates of traffic blockage occurrence were obtained for four time-periods and in different weather conditions: whole days, morning peaks, evening peaks and rainy weather. It was found that for some bus stops, the rate of traffic blockage occurrence could not be calculated. These are stops located on non-public roads, such as on school grounds, or in front of hospitals and thus data on travel speed of ordinary vehicles could not be collected.

The results for two bus stops, that is, Daitakubo bus stop (Ranked as number 2 in the final evaluation) and Kyoiku Center-Mac bus stop (Rank 6; Figure 12) are illustrated. Field surveys confirmed that the Daitakubo bus stop has indeed a high rate of traffic blockage occurrence. The reason is that the stop is located within a short distance between two intersections. Further, there is only one lane in both directions as a result of which the vehicles following behind the bus cannot pass and traffic blockage occurs (Figure 12(a)). Field surveys also confirmed that the Kyoiku Center-Mae bus stop, has a low rate of traffic blockage occurrence even though there is also only one lane in each direction. However, at this stop the bus bay is large enough to allow vehicles to pass or overtake even when a bus stops (Figure 12(b)).

All the results were then arranged in a hospital-type chart for each bus stop. Figure 13 shows an example of such a chart. In discussion with

**a) Daitakubo**

| Rank | Name of Bus Stop | Rate of traffic blockage occurrence | | | |
|---|---|---|---|---|---|
| | | All day | AM Peak 7–8 | PM peak 07–18 | Rainy day |
| 2 | Daitakubo | 0.32 | 0.25 | 0.62 | 0.30 |

**b) Kyoiku Center-Mae**

| Rank | Name of Bus Stop | Rate of traffic blockage occurrence | | | |
|---|---|---|---|---|---|
| | | All day | AM Peak 7–8 | PM peak 17–18 | Rainy day |
| 2 | Kyoiku Center-Mae | 0.04 | 0.03 | 0.00 | 0.44 |

**Fig. 12.** Conditions at Field-surveyed Bus Stops

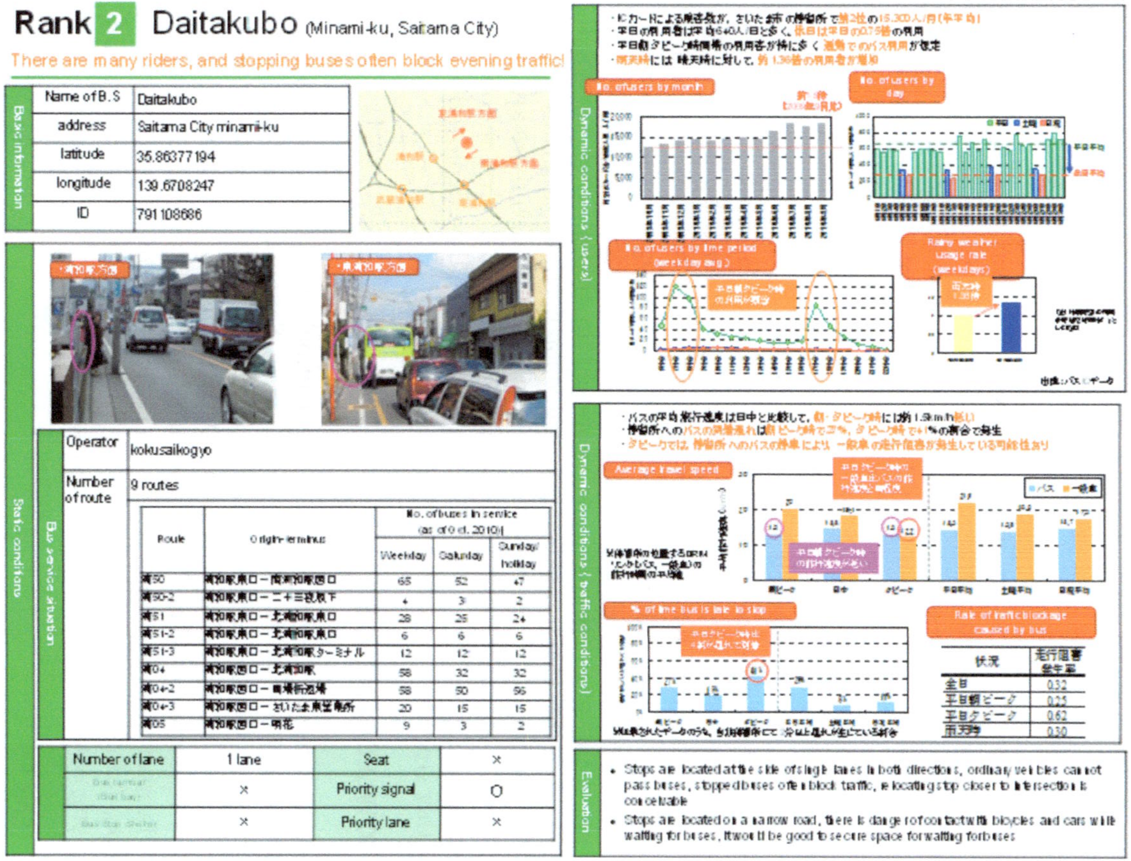


**Fig. 13.** Example of a Bus Stop Chart Created Using the Results of Steps 1-4

the city planners and road administrators, it was confirmed that the information collected was indeed helpful. In fact the road administrators reported that they wanted to use the charts in meetings and in the Transport Strategy Council. (Note that the content of the evaluation of the bus stop in Figure 13 is only an example. This is not the final one used by the administrators.)

## 5. CONCLUSION

A combination of probe car and smart card data was used to study the most urgently required infrastructure improvements in the vicinity of bus stops. The usefulness of the methodology was confirmed by applying it to all bus stops in Saitama City that take payment by smart card. The possibility to extract bus stop traffic blockage points from these two data sources is a new and very practical application. The usefulness of the blockage rate indicator was verified through field surveys.

A second lesson learned from this study was that presentation of analysis results needs to be simple and graphically appealing in order to be useful to the decision makers. It was found that for the purpose of this study, the creation of what in Japan are known as "hospital-style charts" for each bus stop was indeed an attractive way to engage in discussion with road administrators.

Finally, the evaluation criteria proposed in this chapter are highly versatile and the two types of tracking data used for analysis can be collected nationwide in Japan. In the light of this, although there are issues with the use of smart card data, expectations are high that the method proposed here (possibly in slightly changed form) will be deployed widely for further studies.

## ACKNOWLEDGEMENTS

The authors would like to thank the personnel from the Road Planning Division, Department of Land Development, Saitama Prefecture; Traffic Planning Division, Department of Planning and Finance, Saitama Prefecture; City Planning Division, Department of City Development, Saitama Prefecture; Urban Traffic Division, Urban Planning Department, Urban Bureau, Saitama City; and Omiya National Highway Office Planning Division, Kanto Regional Development Bureau for providing the tracking data along with valuable opinions and suggestions.

## REFERENCES

Akiyama, Y., Sengoku, H., Takada, H. and Shibasaki, R. 2011. *Commercial Accumulation Polygon Data Throughout Japan Based on the Digital Classified Telephone Directory*, CUPUM2011, Computers in Urban Planning and Urban Management.

Bagchi, M. and White, P.R. 2004. What role for smart-card data from bus systems? Proceedings of the ICE - Municipal Engineer, Vol. 157, Issue 1. pp. 39-46.

Bagchi, M. and White, P.R. 2005. The potential of public transport smart card data. *Transport Policy*, Vol. 12, 464-474.

City of Niigata, 2007. *Niigata Bus Stop Plan*, 2007 (in Japanese).

City of Saitama, 2011. *Transportation Strategic Plan* (in Japanese).

Hashimoto, H., Mizuki, T. and Takamiya, S. 2014. *Grasping Method of the Bottleneck-Intersections and Influence Area of Congestion Using Probe-Data*. Japan Society of Civil Engineers D3 **70**(5): I_1159-I_1166, (2014) (in Japanese).

Ishida, H., Miura, H., Okamoto, N. and Furuya, F. 2001. *Sampling Rates for Travel Speed Survey with Advanced Information Systems*. Japan Society of Civil Engineers, Vol. 18, pp. 81-88 (in Japanese).

Kawasaki, T. and Hato, E. 2004. *Analysis of Time Space Activity Using Probe Person Data*. Japan Society of Civil Engineers (in Japanese).

Kitano, S., Nakajima, Y., Iryo, T. and Aakakura, Y. 2008. *Longitudinal Analysis for Railway Users Behavior Based on Smart Card Data*. Japan Society of Civil Engineers (in Japanese).

Kinuta, Y., Yabe, T., Nakajima, Y., Makimura, K., Saito, K. and Tanaka, T. 2008. *The Method of Converting Bus Smart Card Data to Travel Time and Bus Trip Data*. Japan Society of Civil Engineers (in Japanese).

Li, D., Miwa, T. and Morikawa, T. Analysis of route choice using private probe data considering heterogeneity in familiarity to OD pairs. *Transportation Research Record: Journal of the Transportation Research Board* (in press).

Makimura, K., Nakamura, T., Chiba, T., Morio, J. and Fuse, T. 2010. *People Flow from Bus Smart Card Data*. Japan Society of Civil Engineers, Vol. 41 (in Japanese).

Momma, T., Hashimoto, H., Matsumoto, S., Mizuki, T. and Uesaka, K. 2011. *Road Traffic Analysis Using Probe Data*. Ministry of Land, Infrastructure, and Transport National Institute for Land and Infrastructure Management (in Japanese).

Nakajima, T., Kitano, S., Kusakabe, T. and Asakura, T. 2009. Empirical Analysis of Behavioral Change of New Railway Line Opening Using Smart Card Data for Automated Fare Payment, Japan Society of Civil Engineers (in Japanese).

Nogami Y., Kataoka, M. and Kumagai, Y. 2011. *Basic Analysis on Public Usage in Kochi Central Area Using IC Card Data* [Deshuka], Japan Society of Civil Engineers, Vol. 44 (in Japanese).

Okamura T., Fujiwara, A., Kanno, M. and Sugie, Y. 2002. *Data Characteristics of Integrated Stored Fare Card System in Urban Public Transport*. Japan Society of Civil Engineers 19, 29-36, 2002 (in Japanese).

Pelletier, M., Trepanier, M. and Morency, C. 2011. Smart card data use in public transit: A literature review. *Transportation Research Part C*, Vol. 19, pp. 557-568.

Tamiya, K. and Seo, T. 2002. *Q-V Configuration for Urban Street Using Probing Car Data*. Japan Society of Civil Engineers (in Japanese)

Trepanier, M., Morency, C. and Blanchette, C. 2009. *Enhancing Household Travel Surveys Using Smart Card Data*. Transportation Research Board 88th Annual Meeting Compendium of Papers.

Yabe, T. and Nakamura, F. 2008. A study on Effect of Smart Card (PASMO) in Tokyo Metropolitan Area, Japan Society of Civil Engineers, 2008 (in Japanese).


## AUTHOR BIOGRAPHY

**Kazuhiko Makimura** (Dr. in Eng.) is Deputy Director at the Institute of Behavioral Sciences (IBS). He has been engaged in various research projects on probe vehicle data (PVD) and smart card data.

**Toshiyuki Nakamura** has been working as an Assistant Professor at the Graduate School of Engineering, Kyoto University since 2011. He has been engaged in research on travel behaviour, especially of bus users and road project evaluation, using data from the 'PASMO' smart card in the Tokyo metropolitan area. Recently his focus has been on analysis of 'LuLuCa' smart card data from Shizuoka, Japan (see Chapter 7).

**Takahiro Ishigami** is Director at IBS for the Urban and Regional Planning Research Division. He has been working on different projects, such as urban and transportation planning, public involvement and road network evaluation. Currently, he is working on practical applications of smart card data analysis for Transportation Planning.

**Ryuichi Imai** (Ph.D.) worked for Nippon Koei Co. Ltd. (2000-2010) before joining the National Institute for Land and Infrastructure Management, Ministry of Land, Infrastructure, Transport and Tourism (2010-2015). Since 2015, he has been Associate Professor in the Dept. of Urban and Civil Engineering at Tokyo City University. He has been engaged in research on road project evaluation by using 'PASMO' smart card data in Saitama prefecture. He is currently also looking into analyses of smart card data from other Japanese public transport operators.

Chapter 13

# Conclusions: Opportunities Provided to Transit Organizations by Automated Data Collection Systems, Challenges and Thoughts for the Future

**N. Wilson[1] and B. Hemily[2,*]**

**ABSTRACT**

Availability of AFC smart card data is part of a broader trend whereby technology is enabling creation of an array of Automated Data Collection Systems for public transportation organizations that will support both off-line service and operations planning, as well as real-time service management and customer information. This wealth of new data sources and analytic tools will assist in enhancing the effectiveness, quality and efficiency of service provided to public transportation customers. However, many challenges exist; some are specific to the use of AFC data including: protection of individual privacy, access and appropriateness of AFC data and ownership of customer data. Other challenges affect all sources of automated data, including: lack of internal resources and technical expertise, conflicting data, corporate data management challenges and lack of support by senior management. Areas for future research are identified, including one area that has been relatively unexplored to date, namely, the use of AFC smart card data to estimate service and especially price elasticities; the latter is an essential tool in developing more creative and sophisticated pricing strategies.

The previous chapters have illustrated how transit smart card data could be used in many ways to understand and develop new insights into passenger behaviour, system performance and policies.

Examples included: developing a better understanding of passenger behaviour through transit origin-destination estimation, route choice and activity behaviour; combining smart card data with other data for exploring trip-making and trip purpose; using smart card data for

[1] MIT. Email: nhmw@mit.edu
[2] Hemily and Associates. Email: brendon.hemily@sympatico.ca
* Corresponding author

calibrating models; and evaluating policy implications related to level of service and equity.

It is clear that the increasing availability of transit smart card data is opening a range of exciting opportunities for research.

However, the real value of mining this rich source of data will be determined by the extent to which it assists public transportation authorities and operators to enhance the effectiveness, quality and efficiency of the services being provided to customers. This chapter broadens the discussion by placing smart card data and its uses into a broader framework for Automated Data Collection Systems (ADCS) and repositions it from the perspective of the transit organizations, both public transportation authorities and operators, who will benefit from the research derived from the use of these data sources. ADCS create many opportunities to support critical transit organization functions, but also face considerable technical and organizational challenges in pursuing these opportunities. The chapter will conclude by exploring one significant unexplored area for research using smart card data and then provide recommendations for the future to build on the efforts to date.

## 1. BACKGROUND

From the earlier days of development of Transit Intelligent Transportation Systems (ITS), there was a recognition among the most progressive transit organizations that the data that would be created by Transit ITS (e.g., Automatic Vehicle Location, Automatic Passenger Counting, Advanced Fare Collection) would be an incredibly valuable resource, which could be used to create information to enhance planning and management and support business processes and decision-making. This can greatly enhance the ability for managers to improve the effectiveness and efficiency of the transit services. Today, one would refer to the notions of Business Intelligence and/or Data-Driven Decision-Making.

A small number of transit systems have carried this belief forward to continuously mine the data resources provided by Transit ITS and these systems are consistently recognised as leaders in the industry. In addition, a number of academic researchers have explored even more refined uses of such data, to create origin-destination maps and a range of analytic tools and analyses. However, the majority of transit systems have not opted, or been able, to pursue this avenue and have rather focused their efforts in using technology to enhance real-time operations and in particular incident management and security; the use of the data created is generally an afterthought. In addition, limited research or guidance is there for transit organizations on how to use this data.

More recently, discussion has increased in the general public realm, as well as within the transportation industry, on related topics including ownership of data, access to data, applications of open data and the potential synergistic benefits derived from data fusion and data mining, popularly known as "Big Data". The transit industry has recently started to explore how it relates to these issues and developments and is increasingly opening its data to researchers and third-party applications developers.

However, the transit industry is seeking to expand its use of data, so the time is opportune to discuss the use of smart card and other data sources for the benefit of the transit industry.

## 2. AUTOMATED DATA COLLECTION SYSTEMS (ADCS)

Previous chapters focused specifically on research based on data derived from smart card systems, but it is important to realize that smart card data is only one of an array of data-generating systems. Most of the Automated Data Collection Systems (ADCS) supporting transit organizations derive from the ITS technologies deployed. ITS is a suite of different systems that are often interrelated. Some might be for real-time activities (e.g., monitoring, control, security), while others are specifically designed to produce data for analysis. However, all of these systems produce data, including logs on events, faults, etc., that form a web of automated data sources.

### 2.1 Automatic Vehicle Location (AVL) Systems

Historically, AVL systems started emerging 30 years ago. On rail systems location is provided using track circuitry, while bus systems today typically use Global Positioning Systems (GPS) as the prime location tool for Computer-Assisted Dispatch/Automatic Vehicle Location (CAD/AVL) systems. The CAD/AVL system is the heart of most Transit ITS deployments. It continuously tracks all transit vehicles in real-time, which enables efficient and effective operational control, incident management, security response and service restoration.

By comparing the real location of vehicles to their scheduled location, it enables continuous monitoring of schedule adherence. This can then be used to provide next stop announcements and to calculate estimated time of arrival of vehicles at all stops downstream and thus drive real-time customer information at displays at stops, on the Internet, on mobile devices, etc.

But AVL systems also provide a wealth of data from on-board devices (e.g. location, door opening sensors, odometer, etc.) that is geo-coded and/or time-stamped describing what the transit vehicles are doing. This in turn could be transformed into information on schedule adherence and On-Time Performance (OTP), running times, dwell times, delays, vehicle speeds, etc. It is important to recognise that it is not only the GPS location that is important, but that the monitoring compares real-time outcomes to schedules.

CAD/AVL systems are typically used to capture information from a number of other on-board sensors (passenger counters, wheelchair ramp, bicycle rack, etc.) that can also provide valuable information.

## 2.2 Automatic Passenger Counting (APC) Systems

APC systems have also been available for two to three decades and are typically based on sensors mounted in doors for buses, which have channelized passenger movements and counters at gates for systems, which have fare barriers. APC data, when properly matched to specific stops, can give detailed profiles of customer activity by stop and time of day. This provides a wealth of information on customer demand. This data is typically not available off the vehicle in real-time.

## 2.3 Automatic Fare Collection (AFC) Systems

Early AFC systems used magnetic technology and generally had extremely limited data collection and processing capabilities. However, over the last decade, AFC technology has been radically transformed. As discussed in previous chapters, AFC systems today are increasingly based on contactless smart cards, tapped at a reader to register the transaction and are geo-coded. In some systems, notably those which have fares differentiated by trip distance or zone, the smart card is tapped both on entry to and exit from the system. AFC data have not typically been available in real-time, but this may become possible in the future.

To date most of these smart cards are issued by the operator or authority. However, Open Systems using contactless bank cards are being introduced and new Mobile Ticketing technologies using optical or near-field-communication-equipped mobile phones are deployed in a growing number of cities around the world. Whatever the technology, the key characteristic is recording geo-coded individual fare transactions, which could be linked to a specific card; this enables the analyses highlighted in previous chapters.

Of these types of systems arguably AFC and AVL are the essential elements to attain most of the benefits achievable from the use of ADCS, especially since AFC can serve as a limited form of APC when combined with AVL data.

## 2.4 Other Pertinent Data Systems

Although AFC, AVL and APC systems are the core sources of automated data, other sources of data exist and are increasingly being considered as sources of information for transit planning and management purposes as more advanced data fusion and mining tools are developed.

These include:

- **Transit Signal Priority (TSP)**

  TSP systems are designed to improve travel times and/or reliability at signalized intersections. First generation systems collected little information and did not permit matching data from the bus and the traffic controller. Newer generation systems should collect data on priority requests and responses and this should enable planners to measure the effectiveness of priority strategies, as well as explore more aggressive strategies.

- **Vehicle Health (Mechanical Alarms)**

  CAD/AVL have always included mechanical alarms to alert control room staff about impending mechanical failures, to take remedial action and alert maintenance staff. While this data has not been extensively used to date, with improved reliability of sensors, this should provide a potentially valuable source of information on vehicle health and enable more advanced fleet monitoring and maintenance planning and management.

- **General Transit Feed Specification (GTFS)**

  GTFS was originally developed as a simple but robust standard to characterize transit routes, stops and schedules that might be used to populate trip planners, such as Google Transit, but also on third-party mobile devices.

  More recently, many researchers and other experts have recognised that GTFS data also provides a remarkably simple way to build transit network models that could be combined with Geographic Information Systems (GIS) or forecasting models for different purposes, including analysis of performance, accessibility, equity, etc. (see for example, Chapters 8 and 11 in this book).

## 3. A CONCEPTUAL FRAMEWORK FOR ADCS IN A TRANSIT ORGANIZATION

It is obvious from the above discussion, that there is a growing array of automated (and other) data sources that are available for off-line and/or real-time use by transit organizations. The following section provides a conceptual framework of the ADCS as it relates to transit organization functions.[3]

### 3.1 ADCS and Key Transit Organization Functions

ADCS have the potential to affect several key functions, which any public transport organization must provide, including both off-line and real-time processes. The distinction between off-line and real-time functions is important both because of the difference in data that is typically available off the vehicle in real-time and because of the difference in computational

[3] This framework is more fully articulated by N. Wilson in the chapter entitled Opportunities Provided by Automated Data Collection Systems, in "Restructuring public transport through Bus Rapid Transit: An international and interdisciplinary perspective", edited by Juan Carlos Munoz and Laurel Paget-Seekins, published by Policy Press (2015).

requirements for real-time applications. The key off-line functions, which could be enhanced by ADCS are service and operations planning and performance measurement. The key real-time functions are service and operations control and management and customer information.

***Service and operations planning*** include specification of services offered as well as basic determinants of efficiency in providing these services, here known as operations planning. Fundamental policy decisions affecting the service offered to the public involve network and route planning, frequency setting and timetable development. Given the underlying modal technology, these decisions largely specify the service characteristics as perceived by the public, which will determine their interest in using the system. The operations planning process focuses on vehicle and crew scheduling, which are key determinants of the cost of operations given the service plan and labour constraints and pay provisions.

ADCS have significant impact on all aspects of service and operations planning, first and foremost through provision of large amounts of data with measurable accuracy. ADCS data is replacing largely manually collected data with its typical connotations of small sample sizes, uncertain and hard-to-measure accuracy and bias. For example, estimation of origin-destination travel patterns previously relied on passenger surveys and used manual passenger counts to expand the resulting seed matrix to the full system ridership. With ADCS systems, as seen, a seed origin-destination matrix reflecting well over half of all passenger journeys could be inferred from ADCS data and then expanded to the full system ridership using the same ADCS data. This should result in more effective service plans and more efficient operations plans, directly as a result of ADCS systems.

***Performance measurement*** is fundamental in assessing all aspects of service delivery. It allows measurement of system performance against policy targets, but is also enabling a more refined measurement of the personal experience of customers.

At the system level, public transport is increasingly expected to deliver service within specified policy-determined quality ranges, often known as Service Standards or Targets. ADCS allow management to measure and report system performance as compared to service standards, and thus ascertain degree of success with respect to promised level of service. This is all the more important if the service is being provided under a contractual relationship between a public organizational authority and a private (or public) operator. The service contract specifies the service targets, performance measures and potential financial incentives/disincentives and the ADCS provides a neutral tool for measuring performance against these targets.

From the customer's perspective, surveys have consistently revealed that service reliability is one of the most important service attributes, but it has been almost impossible in the past to assess service reliability using manually collected data because of the inevitably small sample sizes practical with such labour-intensive data collection methods. Now, with AVL systems, it is practical to amass large numbers of observations, even of a single scheduled vehicle trip, which could be used to support a range of reliability metrics of a traditional operator-oriented nature, for example percentage of trips "on time".

In addition, by combining AFC and AVL data, it is now possible to explore and measure the real experience as perceived by customers. For example, one can measure service reliability for an individual customer by tracking the travel activity of a single card (without of course knowing who that individual is to protect privacy). ADCS enable measurement of service reliability and other attributes in ways not feasible before.

***Service and operations control and management*** deals with day-to-day operations management, in particular responding to unexpected events such as incidents which disrupt normal operations, or significant changes in demand. Depending on the level of the event it might not be feasible to continue to operate the service as planned, at least for a period of time and so an alternative plan might be developed and deployed immediately.

ADCS systems make it possible to respond more effectively to unexpected events, principally through AVL, which provides current locations of all vehicles in the system making it possible to develop a better recovery strategy than without this information. AFC data has the potential to further enhance the response to unexpected events by providing the decision-maker with information on the typical travel patterns near the disruption at this time of the day so that a better strategy could be developed.

***Customer information*** allows the individual customer to be informed of the state of the system, which is particularly important in the case of disruptions and assists them in their travel planning, given deviations from the operations plan. Customers expect current and accurate real-time information at all the stages of their journeys through a variety of media and if public transport is to be perceived as a high quality alternative to driving it must meet these ever-increasing expectations.

ADCS allows targeting of dynamic customer information to the individual through a combination of real-time AVL data and detailed profiles of the travel patterns and preferences of the individual developed through analysis of their historical travel behaviour as revealed through AFC data. Pre-trip information could be based both on the operations plan for advanced trip planning, as is the norm for existing journey planners, or based on the current state of the system for immediate and en route trip planning and re-planning when unexpected events occur. The value of the AFC data cannot be underestimated; for a successful customer information system, only information of value to the individual, given their current (or anticipated) trip-making, should be communicated. To avoid information overload, the customer must be provided only important and pertinent information.

## 3.2 Analytic Framework

The interrelationships between the different transit organization functions and the roles that could be played by ADCS are illustrated in Figure 1. This figure shows the heart of the system which is responsible for integration of the data coming from the ADCS to form a comprehensive picture of the current system state, the analysis of this data to support both the real-time and off-line functions and the prediction of the implications of different strategies on future system performance.

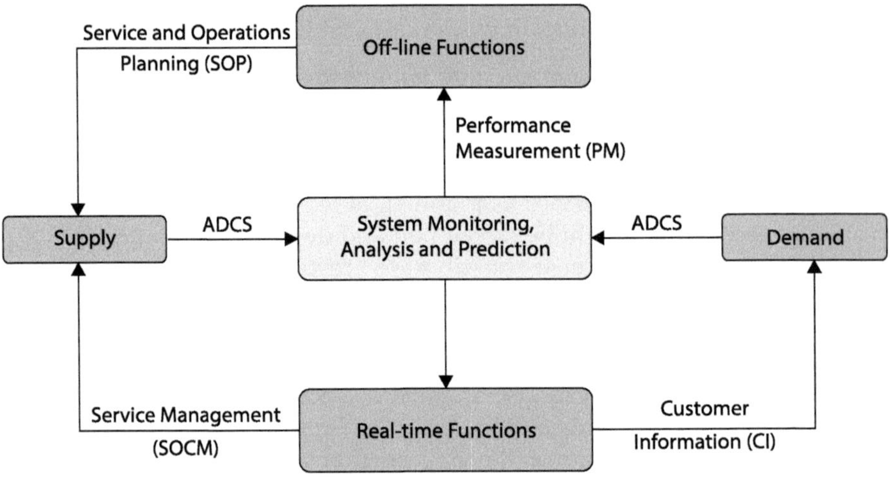


**Fig. 1.** ADCS and transit organization functions

From this figure it is clear that the ADCS, while essential for effective public transport, are just the first step toward optimizing system performance. Analysis methods are required to develop a deep understanding of factors that determine performance. Prediction methods are also essential in order to anticipate the outcomes of particular actions and select preferred strategies. Ultimately the goal is to develop analysis and prediction methods, which can function effectively in real-time to support the supply management and dynamic customer information functions. In the short-term, if the computational burden is too high for real-time application, significant value might be achievable through the planning and performance monitoring functions.

Given the complexity of predicting performance of public transport systems, which involves understanding customer behaviour as well as developing both short-term and longer-term service and operations plans, the analysis methods required will inevitably be complicated. They will certainly include simulation-based performance models, which are the only credible way to join both customer response to information and decision support for operations controllers and managers. Their development will be a demanding research activity, which will need a deep understanding of both the demand for transport services and their performance.

While a comprehensive model encompassing all these desirable features remains in the future, there has been progress on some of the key modules and analyses required for such a model. The contributions in this book illustrated the wide range of research underway around the world, which are leading to new analyses of system performance and customer behaviour, as well as to the development of new methodological tools that may someday be incorporated into the above analytical framework.

For example, smart card data is being used to research:

- Path choice/transit assignment (including impact of transfers, network choices, crowding, information, etc.).
- Transfer patterns.
- Route/vehicle loading.
- Service reliability as experienced by customers.
- Variations by time of day, day of week, etc.
- Inference of residence location and socio-demographic information.
- Comparison with travel surveys to perform validity checks.
- Customer retention rates.
- Impact of weather on travel behaviour.
- Shopping vs. mode access behaviour.

Analysis of smart card data is helping to formulate potential real-time operational management modules, such as:

- Real-time changes to operational plans,
- Real-time intermodal coordination, and
- Incident management inputs/outputs:
  - Likely scenarios for traveller response,
  - Contingency plans,
  - Emergency information provision,
  - Transfer management, etc.

It is clear that this research has many practical applications for transit organizations, but the challenges in transferring this knowledge and in building advanced analytic tools within transit organizations, are in most cases significant.

# 4. CHALLENGES

This section will discuss some of the technical and organizational challenges that create barriers to transferring the knowledge gained to transit organizations so that they can enhance the effectiveness, quality and efficiency of service provided to transit customers. These are broadly defined and based on extensive discussions conducted with transit organizations, but will not necessarily be those experienced by any given transit organization.

## 4.1 Challenges Specific to AFC Data

The previous chapters outlined many of the complex methodological challenges met in using smart card data for research. Some of the methodological challenges encountered include:

- Data quality.
- Large volume of data produced and ability to process.
- Methodologies to expand data samples.
- Determination of geographical location, especially in open systems without check out.
- Distortions of behaviour caused by pricing.
- Distortions in longitudinal series caused by card expiry date, etc.

Transit organizations can also face significant policy or organizational challenges to use smart card data. Some of these include the following.

***Protection of individual privacy*** is a paramount societal policy of special concern that affects the use of smart card data. Rules exist at different levels, both national/state or province/regional, and can vary significantly from jurisdiction to another. In some case, efforts to anonymize cards might be insufficient to satisfy some privacy advocates and policymakers.

***Access and appropriateness of AFC data*** has been a serious limitation in past AFC technologies. Fare collection technologies have been traditionally designed to control the collection of revenues and ensure financial accountability. From this perspective, revenues must be counted and secured, but ridership need only be monitored at broad aggregate levels (e.g., by bus by day and perhaps by run); they were not intended to collect stop-level passenger data. This is changing rapidly when introducing new systems, but legacy smart card systems will not necessarily have each transaction logged and geo-coded. Much of the research illustrated in previous chapters derived from recent advanced AFC systems that enable time and geography-sensitive customer-level monitoring.

***Ownership of customer data*** is always an issue, but will be even more complex as new approaches to fare collection involving third parties (e.g., banks, mobile device companies) are deployed. Transit organizations often neglect to carefully specify the public ownership of data in systems that are primarily designed for purposes other than to collect data, but this will become even more complex and important, in a future involving Open Fare Collection System and Mobile Ticketing. Private companies, such as banks and mobile communication carriers, are more likely to be aware of the importance of the ownership of data and to have the required expertise in the associated legal aspects. Public entities will have to significantly expand their expertise in this area if they intend to retain the ability to use the data created by AFC systems.

## 4.2 Other Challenges Related to ADCS (including AFC data)

Beyond the specific challenges revolving around the use of AFC data, there are many other significant challenges related to the effective use of ADCS in transit organizations,[4] including:

- Lack of internal resources and technical expertise,
- Conflicting data,
- Corporate data management challenges, and
- Lack of support by senior management.

***Lack of internal resources and technical expertise***: In most agencies there is a lack of resources and technical expertise for analysis using ADCS data. This requires expertise on one hand on technical tools and processes for data mining and visualization, but on the other, on transit business processes. At the same time, there is generally a lack of resources for Information Technology (IT) data management support.

***Conflicting data:*** Conflicts sometimes occur between different sources of data, which can undermine credibility and dampen use. Problems encountered include:

- Lack of an integrated data warehouse and the resulting existence of multiple databases with different coding of the same information (e.g., bus stop inventory),
- Multiple sources of GPS location, from different on-board systems (e.g., AVL vs. AFC),
- Conflicting ridership data from different sources, such as APC and AFC systems.

***Corporate data management challenges:*** There are also various challenges related to organization of automated data within the organization and its

[4] This section is based on research by B. Hemily on behalf of the U.S. Department of Transportation and ITS America, entitled The Use of Transit ITS Data for Planning and Management and Its Challenges; a Discussion Paper, Final Report – Revised July 28, 2015.

management. In many transit organizations, the IT Department might be under-resourced and data management will be a secondary priority compared to basic IT network hardware and software responsibilities.

Some of the typical data management challenges that have been identified include the following:

- ITS technology supplier ownership of data in legacy systems, limiting use by transit organization,
- Data storage: managing the volume of data, especially if there is a lack of an integrated data warehouse,
- Lack of (or unclear) data retention policies,
- Lack of systematic inventory of databases,
- Use of proprietary, or just different, data formats and even definitions by the suppliers of ITS technologies, making interoperability and data integration challenging,
- Missing or corrupted data (including "Bad Day" anomalies),
- Lack of diagnostic tools provided by suppliers to determine cause of data collection/matching failures,
- Lack of clarity about policies and procedures with respect to the management and provision of Open Data, etc.

***Lack of support by senior management:*** More generally, policy boards and senior managers of transit organizations need to continuously focus on ensuring sufficient funding to operate and expand the transit system and building the stakeholder coalition to do so. Technology is often a secondary concern and they are often not very interested in ITS, even less so in the data that ITS create. The transit industry is by-and-large characterized more by an operations-driven culture than by a data-driven decision-making culture. However, it could be observed that interest in ADCS and the use of data for management and policy is growing, creating more opportunities for fruitful collaboration between academic researchers and transit organizations, as illustrated by some of the examples in previous chapters.

## 5. AN UNEXPLORED AREA FOR RESEARCH USING SMART CARD DATA: ELASTICITIES AND PRICING STRATEGY

One area of research has remained relatively unexplored to date and that is to use AFC smart card data as a tool for measuring customer sensitivity to service and price changes. This is by calculating the related elasticity, i.e. the percentage change in ridership to the related percentage change in service supply (service elasticity) or price (price elasticity).

Smart card data allows longitudinal analysis of each customer by monitoring trip-making of each card, as identified through their unique card number (without identifying the specific individual). This means that changes in ridership could be corelated with changes in service or fare levels, providing an obvious source of data for calculating elasticities. This analysis could be further segmented: by fare category represented by the card (adult, student, senior); by type of rider (as represented by the fare product they use, e.g., pay as you go for occasional riders and monthly or annual passes for frequent riders); by geographic area; by trip purpose (commuter, school, shopping); etc. Elasticities are very hard to measure manually and the last significant research in this area dates from several decades ago.[5]

Service elasticities would be valuable, but calculating fare elasticities is perhaps even more important for transit organizations, since they directly affect the organization's pricing strategy and thus the "Demand" for transit service, but are areas of much uncertainty. Figure 2 illustrates how "Pricing Strategy" relates to the earlier ADCS conceptual framework.

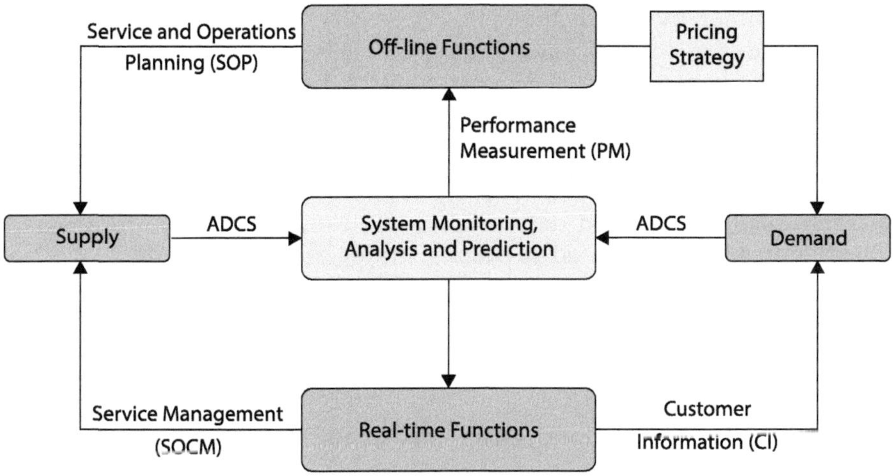


**Fig. 2.** "Pricing Strategy" as a new ADCS-related transit organization function

The pricing strategy of a transit organization affects the heart of revenue management and is a critical function. Introduction of smart card technology was often promoted as enabling greater flexibility in the pricing strategy: new products could be much more easily introduced and creative targeted or time-limited fare products could be experimented with. However, the risks in experimenting with revenue management are huge and the uncertainty has been great. To date there has been:

[5] TCRP Report Volume 95 Chapter 12 (2004) Traveler Response to Transportation System Changes synthesizes much of the prior research on fare elasticity values.

- Little accessible information on smart card use, customer behaviour and impacts to help in the planning of new smart card deployments, and
- There had been no information available to transit organizations from previous experience with AFC technology on the behavioural impacts that might result from introducing smart cards, such as potential customer switching between media, changes in ridership patterns, etc.

As a result, there was little basis for managing the associated potential risks. Given the requirement for transit organizations to be conservative with the stewardship of public funds, there is little incentive to innovate fare policy, even when introducing new, more flexible, AFC technology.

However, the data that is becoming available from existing smart card systems is providing a valuable resource that could help transit organizations better understand the revenue risk vs. the ridership potential of new pricing strategies. This is the essential question that transit organizations must ask themselves, with the important corollary of understanding how any new pricing strategy or product affects equity, by type of customer, by jurisdiction, etc.

Although this analysis extends beyond the normal realm of engineering and planning researchers, it is important to transit organizations, could be analysed through AFC smart card data and is part of the global ADCS conceptual framework. In addition, limited targeted pricing innovations could be structured and tested and then monitored using smart card data.

Research using smart card data might help transit organizations answer a range of uncertainties related to pricing strategies:

- What is the pass multiplier (Monthly, Weekly, Daily)?
- How sensitive are customers to price increments per zone?
- Is there a market for special fares for short trips?
- What might be the maximum allowed time for a journey?
- What is the sensitivity to peak vs. off-peak pricing strategies?

There are many other suggestions for innovative pricing strategies where analysis would help and might be feasible to explore using AFC smart card data. These include:

- Evening and/or weekend fare.
- Weekend pass.
- University or Employer based discounted annual pass (U-Pass, Eco-pass) use rates by time of the day.
- Summer pass for students.
- Student freedom pass (after 4PM/weekends).

- Co-pricing with sports/entertainment events.
- Passes to condominium buyers (in lieu of parking).
- Social fares (unemployed).
- Loyalty schemes.
- Shared-use mobility co-pricing (bike-share, car-share), etc.

## 6. CONCLUSIONS: LOOKING TO THE FUTURE

This book has presented many examples of the exciting research underway that is building upon the growing availability of AFC smart card data, thus illustrating its value as a resource to analyse important issues related to transit system performance, customer behaviour and even public transportation policy issues.

This chapter has shown that availability of AFC smart card data is part of a broader trend whereby technology is enabling creation of an array of Automated Data Collection Systems that will support both off-line service and operations planning, as well as real-time service management and customer information. This wealth of new data sources and analytic tools will assist transit organization to enhance the effectiveness, quality and efficiency of service provided to customers.

This chapter has also identified one area that has been relatively unexplored to date and that could benefit from more in-depth research using AFC smart card data, namely, the analysis of service and especially price elasticity that are an essential tool in developing more innovative and sophisticated pricing strategies.

Looking towards the future, the following are some recommendations to build on the efforts to date.

***Methodological Research***: The research described in this book has shown the progress made in addressing substantial methodological issues such as the inference logic required to build Origin-Destination matrices from open system smart card data. Nonetheless, many methodological issues still remain. Some of these are generic in nature, while others are unique to the AFC architecture and pricing strategy in a specific community. More research will stimulate more discussion around key issues to build consensus within the analytic community.

***Data Fusion***: This book has already illustrated examples of research based on data fusion of smart card data with other sources of data. One area that merits more attention might be efforts to combine smart card and demographic/socioeconomic data to define cohesive market segments as a basis for analysing travel behaviour. Privacy concerns typically prevent direct knowledge of an individual, but fare categories give a first cut at segmentation and might be combined with other sources of data.

***Price and Service Elasticity Research***: As mentioned above, there is a unique opportunity offered through AFC smart card data to research customer sensitivity to service and price changes and thereby measure service and price elasticity for different market segments. These will be of particular importance for the transit organization's pricing strategy.

***Technology Transfer***: One of the exciting aspects of the body of research emerging from AFC smart card data is that it relates more directly to the needs of transit organizations than do other areas of research. It is often more directly accessible and applicable for transit organizations. For example, few transit agencies have advanced demand estimation or mode choice models, but all transit organizations check on-time performance and service reliability, even if only manually. This provides researchers with an ongoing basis for dialogue with transit organizations: they need access to the smart card data, but can offer as a *quid pro quo* analyses that are pertinent to transit organizations. This can serve to bridge the gap that often exists between the research and practitioner communities and much of the research in this book illustrates the kind of partnership that can emerge from such exchanges.

***ADCS Capacity Building***: However, as outlined before, transit organizations face many significant challenges in using AFC and other ADCS data. There is a clear need for transit organizations to build their ADCS capacity. This means addressing the organizational and data management challenges, developing the tools, resources and ability to transform data and analyses into actionable information, but mostly building the business case that will convince senior management and policy boards of the value of data-driven decision-making and the positive return on investment in building and supporting the systems that will create and analyse the required data. The research community should work with transit organizations in developing these business cases and in building this capacity.

***Towards Big Data and Smart Cities***: With expansion of ADCS data sources internally within transit organizations and the universal growth of open data sources, more avenues should open up for exciting research. This is leading to the much talked-about world of Big Data and Smart Cities. AFC smart card data may actually become one of the pillars to pursue these visions. Beyond data fusion, development of the data mining methodologies will be a focus area of growing importance in this respect and researchers of AFC smart card data are among the pioneers of Big Data.

## AUTHOR BIOGRAPHY

**Nigel Wilson** is Professor of Civil and Environmental Engineering at the Massachusetts Institute of Technology. He leads a public transport research program which features long-term collaborations with leading international agencies including Transport for London, the Massachusetts Bay Transportation Authority (Boston) and MTR (Hong Kong). A major focus of these collaborations is the use of smart card data, with other automatic data collection systems, to improve the planning, control and performance of public transport systems.

**Brendon Hemily** is an independent public transportation consultant focusing on best practices, innovation and the strategic use of advanced technology. He has provided support to the U.S. Department of Transportation and ITS America related to Transit ITS technologies. He previously worked for the Canadian Urban Transit Association and is Chair of the TRB Stranding Committee on Public Transportation Planning and Development.

# Index

## P



## R



## S



## T



## U



## V

An environmentally friendly book printed and bound in England by www.printondemand-worldwide.com

PEFC Certified

This product is from sustainably managed forests and controlled sources

www.pefc.org

PEFC/16-33-415

This book is made of chain-of-custody materials; FSC materials for the cover and PEFC materials for the text pages.